The Montana Mathematics Enthusiast

2007 Monograph 3

Festschrift in Honor of Günter Törner's 60th Birthday

The Montana Mathematics Enthusiast

2007 Monograph 3

Festschrift in Honor of Günter Törner's 60th Birthday

Edited by

Bharath Sriraman

The University of Montana

INFORMATION AGE PUBLISHING, INC.
Charlotte, NC • www.infoagepub.com

Library of Congress Cataloging-in-Publication Data

Beliefs and mathematics : festschrift in honor of Guenter Toerner's 60th birthday / editor, Bharath Sriraman.
p. cm. – (The Montana mathematics enthusiast: monograph series in mathematics education)
Text in English and German.
Includes bibliographical references.
ISBN 978-1-59311-868-6 (pbk.) – ISBN 978-1-59311-869-3 (hardcover)
1. Toerner, Guenter, 1947- 2. Mathematics–Study and teaching. 3. Belief and doubt. 4. Mathematics–Philosophy. I. Tvrner, Glnter, 1947- II. Sriraman, Bharath.
QA11.2.B452 2007
510–dc22 2007043956

Permission to photocopy, microform, and distribute print or electronic copies may be obtained from:

Bharath Sriraman, Ph.D.
Editor, *The Montana Mathematics Enthusiast*
The University of Montana
Missoula, MT 59812
Email: sriramanb@mso.umt.edu
(406) 243-6714

Cover photo: Günter Törner at his residence in Bottrop, Germany, January 2006.

Printed in the United States of America

This commemorative monograph
of The Montana Mathematics Enthusiast
celebrates the contributions of Günter Törner to mathematics
and mathematics education.

CONTENTS

GÜNTER TÖRNER

A True Academic, Friend, Wunderbarer Mensch

Bharath Sriraman
The University of Montana

Günter Törner and Bharath Sriraman, Bottrop, January 2007

Günter Törner (1947–) turned 60 this year on July 29. The idea for this Festschrift spawned three months before his birthday. It is a German tradition to celebrate a scholar's 60th birthday by either organizing a symposium and/or a Festschrift in which colleagues contribute papers related to or stemming out of the contributions made by the academic. In this case, I am

The Montana Mathematics Enthusiast, pages ix–XXX

glad to have to have taken the initiative to compile, edit and organize this collection of articles which comprise this Festschrift to honor my colleague and dear friend Günter Törner, and more importantly surprising him with this book released as part of the monograph series of *The Montana Mathematics Enthusiast* as a (belated) birthday gift during his visit to Missoula and The University of Montana this Fall (Sept. 24–Oct. 12, 2007).

It is appropriate that this Festschrift appear in a journal which has the words "Mathematics Enthusiast" in its title because Günter belongs to the rare breed of academics who makes original and sustained research contributions to both pure and applied mathematics as well as mathematics education. Günter's doctoral dissertation was in the area of geometry, namely the *Hjelmslev planes* with a detailed study of the associated rings which arise from co-ordinatizing these planes. The dissertation was supervised by Benno Artmann, who is himself another mathematics enthusiast, now actively involved in the study of mathematics and art in addition to 4-dimensional geometry. Günter's work was awarded the prize for the best doctoral dissertation at the University of Giessen in 1974, and four years later at the age of 31, he achieved the remarkable feat of being conferred a Full Professorship at the University of Duisburg (now Duisburg-Essen), where he has been ever since.

The Festschrift opens with an artful paper by Benno Artmann himself on the projective plane, and consists among others, of numerous contributions in the mathematics education domain of beliefs, an area which Günter is considered an expert both in and outside Germany. These papers which connect to Günter's research interests over the last three decades have been written by well wishing and distinguished colleagues whose lives, careers and intellectual works have intersected with that of Günter. It took a great deal of time and effort on the part of these colleagues to prepare an article for this Festschrift given the extremely short notice and tight time schedule to bring this to completion. I could not have had this monograph ready without their help. My heartfelt thanks and appreciation go to all the authors in this issue.[1] Please note that I have strived for uniformity to the maximum extent possible which was an exercise in patience and perseverance and at times seemed impossible given the different formats in which the manuscripts arrived (MS-Word, Word Perfect, LaTeX, etc.)

I have been fortunate to know Günter as a friend as well as having worked with him quite intensively during the last several years on numerous projects. The three most valuable things I have learned from him and emulated are: (1) to never refuse help to those who need/seek help, particularly to colleagues that are just entering the field, (2) the art of juggling multiple research interests and academic commitments without compromising one's family obligations, and (3) consistently aiming for high standards of academic service and scholarship of value and interest to the field.

So, joining the authors who honor Günter in their articles and who have made this Festschrift possible, I say *Nachträglich alles Gute zum 60. Geburtstag!*

NOTE

1. Among the colleagues who join us in honoring Günter but were regrettably unable to contribute articles for this monograph are Rolf Biehler (Bielefeld, Germany) and George Philipou (Cyprus).

CHAPTER 1

PICTURES OF THE PROJECTIVE PLANE

Benno Artmann
Goettingen, Germany

Dedicated to Günter Törner on the occasion of his 60th birthday

Günter Törner started his mathematical career with a doctoral dissertation on *Hjelmslev planes.* These planes were named after the Danish mathematician Johannes Hjelmslev (1873–1950), who defined them as a sort of "natural geometries" meaning that distinct lines may meet in more than one point, like it may happen in a real drawing. There is a natural homomorphism $H \rightarrow P$ from a Hjelmslev-plane H onto an ordinary projective plane. The geometric axioms for H are in the "desarguesian case" such that H may be co-ordinatized by a ring that has a unique chain of ideals. Toener went on to study these rings in detail, which today is the main topic of his mathematical research beside his activities in mathematical education and his service to the Deutsche Mathematiker Vereinigung. On top of all this he is interested in the relations of mathematics and art and has even organized meetings on this topic. That is why I have decided to offer him some remarks on the pictorial history of the ordinary (real) projective plane P. I am sorry that I do not know of any nice pictures of Hjelmslev planes or their uniserial rings of coordinates.

The Montana Mathematics Enthusiast, pages 1–12

PICTURES OF THE PROJECTIVE PLANE

I will just line out the historical steps towards pictorial representations of the projective plane, providing historical pictures whenever possible. Unless stated otherwise, drawings are by the author. Ultimately the projective plane stems from the study of perspective in the Italian Renaissance. In our terms, this means to direct ones attention to the lines of sight emanating from the eye of the painter. The lines of sight may be cut with a sphere surrounding the eye and each line gives rise to the identification of two antipodal points of the sphere.

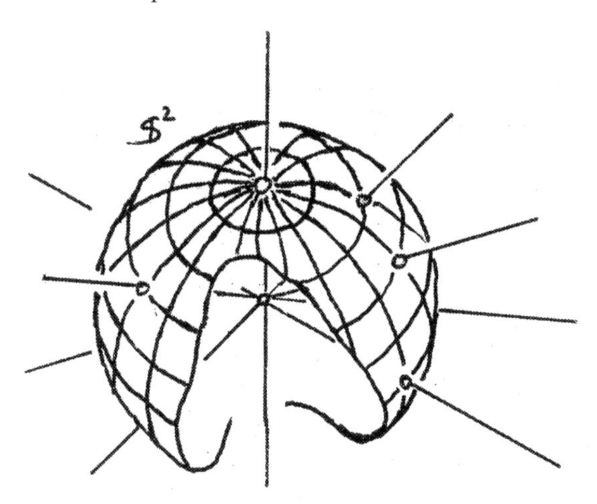

Figure 1.1 Sphere with lines of sight.

Jakob Steiner

In the first half of the nineteenth century, projective geometry came to the foreground of mathematical interests. One of the protagonists of synthetic projective geometry was the swiss mathematician Jakob Steiner (1796–1863), who found the first picture of the projective plane. It is what we today see as an embedding into 3-space (or maybe better into 4-space). He defined the mapping of the two-sphere S^2 into R^3 by

$$(x,y,z) \rightarrow (yz, xz, xy).$$

Since he had found this surface during a stay in Rome, Steiner called it "my Roman surface," but did not publish it. He told his friends Weierstrass and Kummer about it, who published it posthumously. In the year of Steiner's death, Kummer presented the first plaster model. Because the mapping

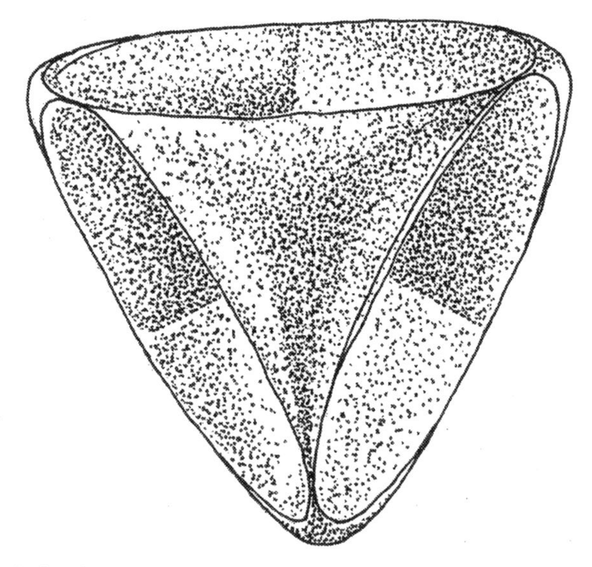

Figure 1.2 Steiner's Roman Surface[1]

identifies antipodal points of the sphere, the model gives us a picture of the projective plane. Steiner himself studied it in the connection with conic sections.

August Moebius

The second mathematician who found a picture of P in quite a different context was August F. Moebius (1790–1868). In 1858–65, he was looking for a way to determine the volume of a polyhedron from its (triangulated) surface. Before him, people had found ways to speak meaningful of the area of a closed plane polygon. Moebius realized that here was no way to generalize this to polyhedra. From the local aspects of the surface one can not tell if it contains a volume at all. His famous "Moebius strip" could be part of the surface and hence there was no way of speaking of an inside or outside of the surface (Moebius p. 482–485).

The minimal example presented by Moebius goes like this: First he defines his strip by a sequence of 5 triangles

ABC, BCD, CDE, DEA, EAB,

to be identified along corresponding sides.

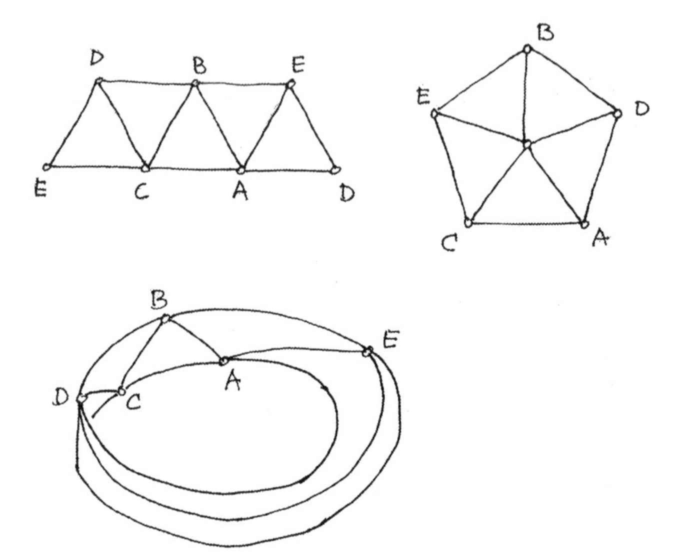

Figure 1.3 Moebius' strip and lid.

The strip is not a closed surface, so he takes a point F (outside the strip) and defines a "lid" by 5 more triangles

$$\text{FAC, FBD, FCE, FDA, FEB,}$$

to be glued to the corresponding sides of the strip. Now there is no free edge left and the result is a closed (non-orientable) surface *M*. Here Moebius did not provide a picture. We, however, may try and see the self-intersection of *M* along the line FX in Fig. 4. Interested readers may try for themselves to put this triangulation on a picture of a cross-cap projective plane as defined in the next section. If we count the faces f, the edges e, and the vertices v of *M*, we get the Euler characteristics

$$f - e + v = 10 - 15 + 6 = 1$$

of the projective plane.

Interlude

Before going on we will have a modern look back at Moebius' construction and the ordinary topological representation of *P* as a (sphere with)

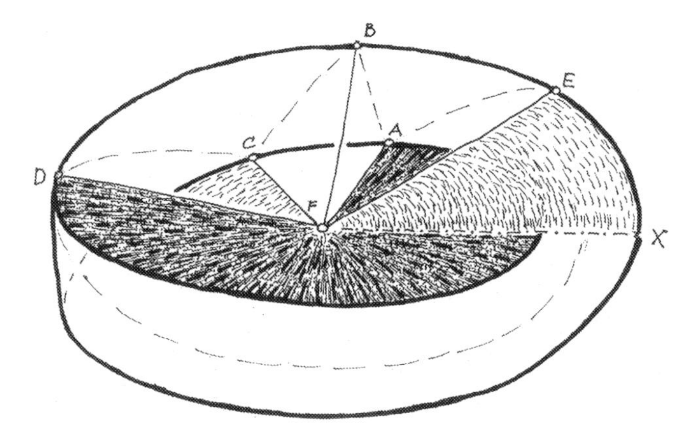

Figure 1.4 The projective plane after Moebius, the self intersecting triangles FAC and DFE are shaded.

cross cap. What surface results from identifying antipodal points of the sphere *S*? Instead of identifying, we may select one of the two antipodal point and delete the other one. Proceeding thus, we split the sphere into two polar lids and a central belt.

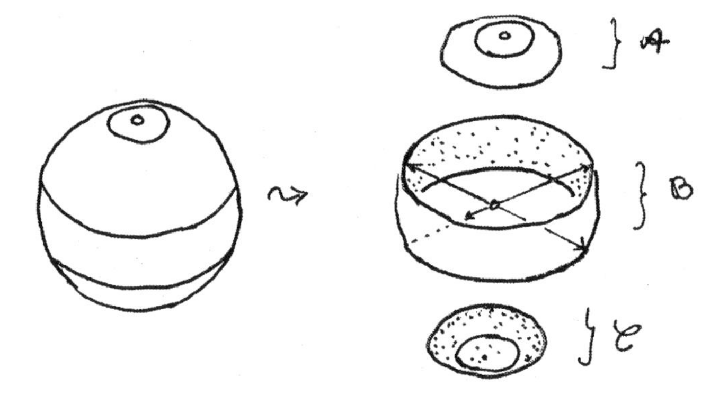

Figure 1.5

Of the two lids, we delete the "southern" one *C* and keep *A*. Of the belt we still have to throw away one half, but have to be careful about identifying the ends. This way the belt becomes a Moebius strip *M* and the lid *A* has to be glued along the border of *M*. If we subdivide strip and lid into the triangles of Moebius (with the center of *A* marked F), then we are just back in his situation.

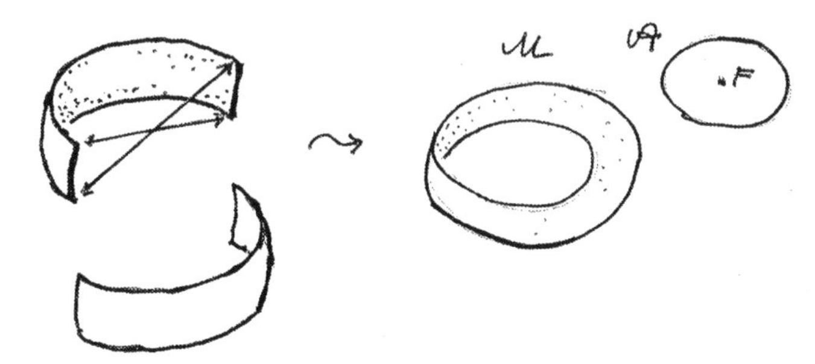

Figure 1.6 Moebius strip and lid.

The second more familiar way of representing *P* is by a cross-cap: throw away the "northern" hemisphere but take care of identifying opposite points on the equator right.

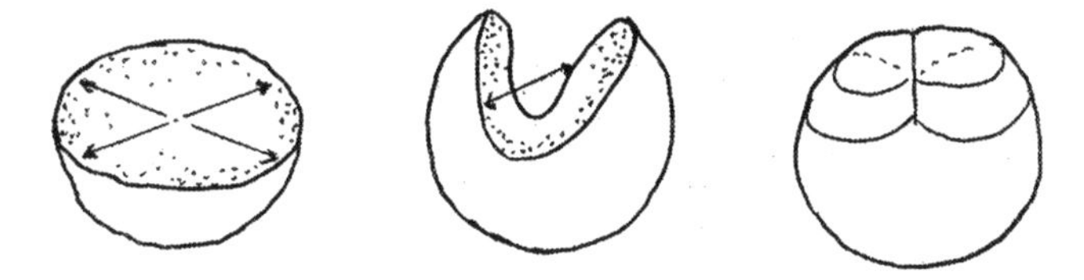

Figure 1.7

Werner Boy

Werner Boy[2] (1879–unknown) was a doctoral student of Hilbert, his published dissertation (Boy 1903) is about the (differential) topology of closed surfaces. He points out that the most well known "one sided" surface is the Klein bottle, shows it and points out that its (Euler-) characteristic is 0.

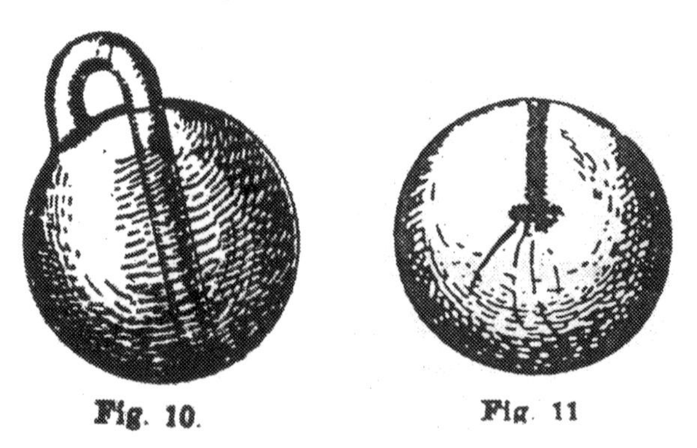

Figure 1.8 The Klein bottle, Boy's fig 10 and cross cap, his fig 11.

Next he mentions Steiner's Roman surface as an example of characteristic 1. Then he proceeds to a new picture, which is just our cross cap. Because he does not mention any earlier picture of the cross cap, he may have considered it to be well known. In fact, there is an earlier sketch of it and some more non-orientable surfaces by W. Dyck (1856–1934) in his (1888, figures in the appendix).

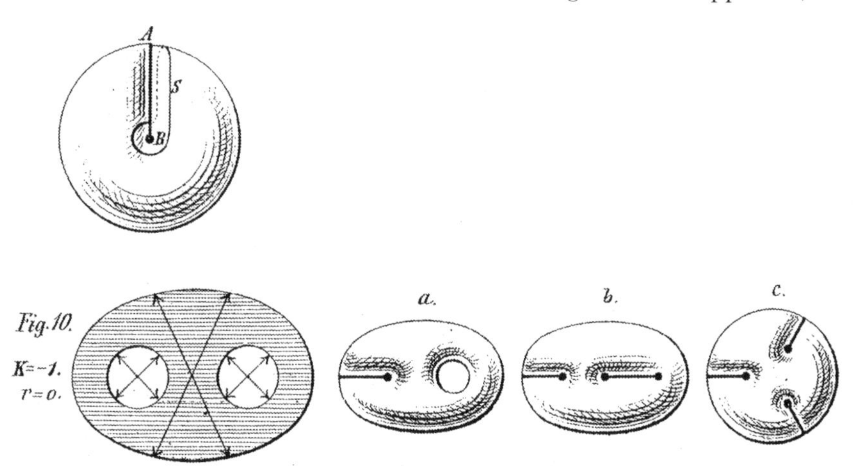

Figure 1.9 Cross cap and other surfaces by Dyck 1888, plate 2.

Unfortunately, the cross-cap has two pinch-points where it is not differentiable. Boy proceeds to construct an everywhere differentiable realization of *P*, the famous "Boy's surface" and shows pictures of it from a front and reverse side:

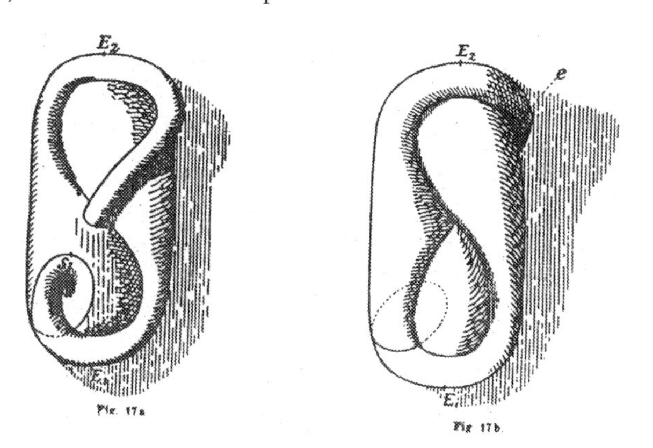

Figure 1.10 Boy's figures 17 a,b

Plaster models of this can be found in the collections of many mathematical institutes, excellent photos of these models and related ones are in (Fischer I, 1986, p. 110–115). Boy himself draws still another view:

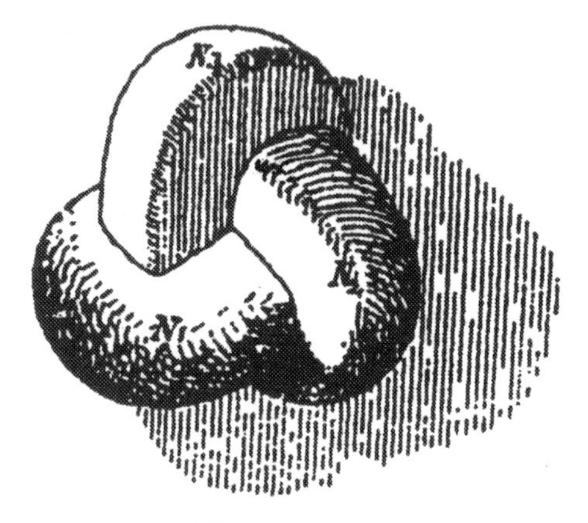

Fig. 20 a.

Figure 1.11 Boy's figure 20a.

From the pictures like Boy 17 a,b it is somewhat hard to see the internal structure of the surface. As a remedy for this, George Francis (1980 and 1987) had the great idea to cut suitable "windows" into the surface, and that way it is easy to see what is going on "inside."

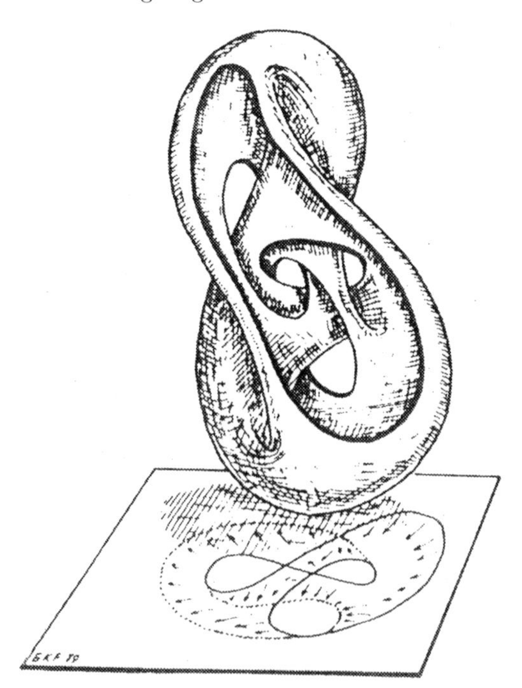

Figure 1.12 Boy's surface with 4 windows, G. Francis 1980, p. 203.

From Mathematics to Art

Here it is in order first to remember some statements concerning principal aspects of the relations of pictures in the sciences and the arts by the art-historian and philosopher G. Boehm (2001, 53):

> Scientific pictures have their meaning outside themselves. In other words they are constructed to show something outside themselves, they are instruments, in contrast to works of art, which have their meaning in themselves. Scientific pictures intend a unique meaning, whereas works of art admit of different interpretations. Richness of metaphors typical for works of art is not useful in scientific pictures.

On the other hand, Boehm stresses that with the passage of time scientific pictures may loose their immediate purpose, and in the process of aging may just keep their aesthetic value and hence are transformed into works of art. Look at Boy's and Dyck's pictures Fig. 10 under this aspect! That means, in some senses at least, mathematical instruction is reconciled with aesthetical attraction. Keeping these considerations in mind we finally proceed to some works of art inspired by various forms of the projective plane, now just representing themselves and serving no other purpose.

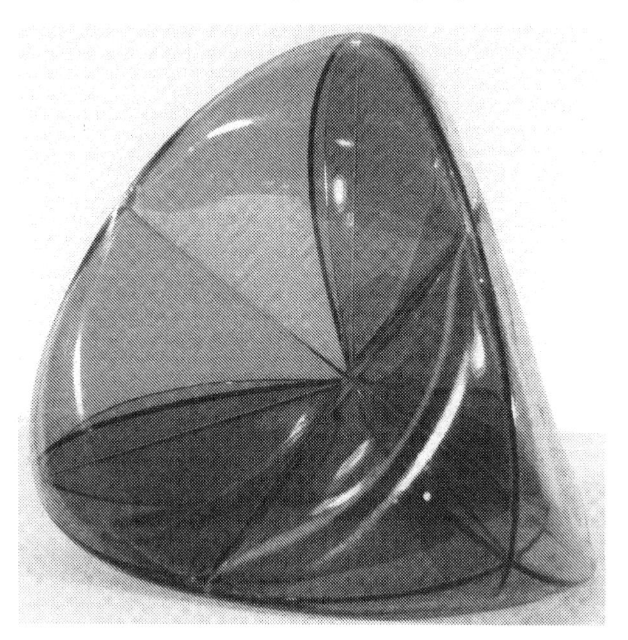

Figure 1.13 Ruth Vollmer: Steiner's Roman Surface 1970.[3] Diam. 30.5 cm, Collection Dorothea and Leo Rabkin, New York.

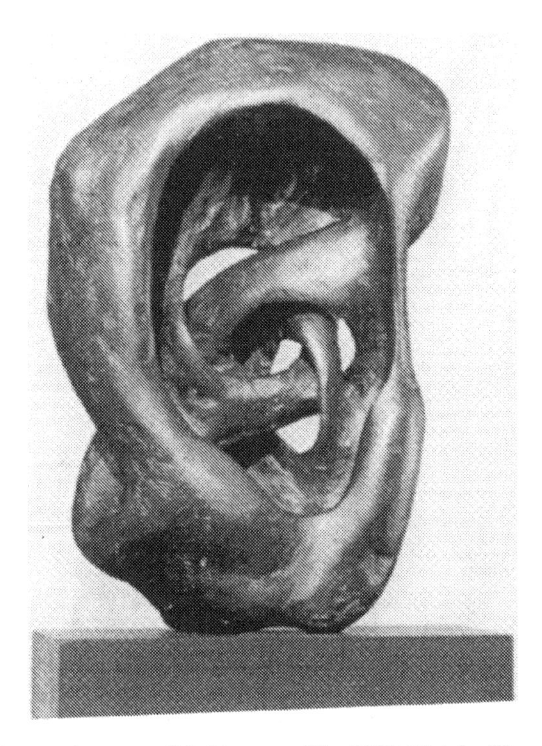

Figure 1.14 Benno Artmann: Ich bin ganz Ohr 1982. Height 38 cm (Photo B. Artmann). Again Boy's surface with 4 windows, inspired by G. Francis.

Figure 1.15 Boy's surface[4] in Oberwolfach (Photo B. Artmann).

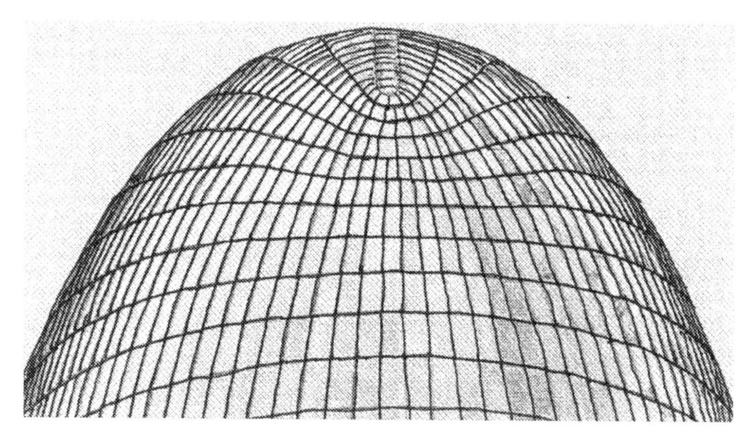

Figure 1.16 The modern building of Peek and Cloppenburg Köln/Cologne 2005. Renzo Piano architects. Drawing after a photo by Fechner/Koch.

NOTES

1. For historical and modern details see:
   Steiner Werke II, Nr. 32, Kummer 1863, Hilbert an Cohn Vossen 1932 Appendix p. 300/301, for excellent pictures of a plaster model Fischer 1986 I and for more details Pinkall in Fischer 1986 II, 108/109.
2. Apparently the last thing that is known about Werner Boy is, that he was a Gymnasium teacher at a school in Krefeld in 1908. The author is in contact with the people in the archives in Krefeld and hopes to find out more about Boy's later life.
3. Photo Oliver Klasen. Copyright ZKM Karlsruhe. With kind permission of the Zentrum fuer Kunst und Medien Karlsruhe, Germany. (p. 98 of Rottner-Weibel 2006).
4. The Boy-surface at Oberwolfach was conceived by Ulrich Pinkall in Berlin

REFERENCES

Boehm, Gottfried: Zwischen Auge und Hand. p. 43–53 in: B. Heintz and J. Huber, ed.: *Mit dem Auge denken.* Wien: Springer 2001.

Boy, Werner: Ueber die Curvatura integra und die Topologie geschlossener Flaechen. *Math. Annalen 57* (1903), 151–184.

Dyck, Walter: Beiträge zur Analysis situs. *Math. Annalen 32* (1888), 457–512.

Fischer, Gerd: *Mathematical Models* (I Photos, II Commentary). Braunschweig: Vieweg 1986.

Francis, George: *A Topological Picture Book.* New York: Springer 1987.

Francis, George and Bernard Morin: Eversions of the Sphere. *The Mathematical Intelligencer* 2(4) 1980, 200–203.

Hilbert, David and Stefan Cohn-Vossen: *Anschauliche Geometrie.* Berlin, Springer 1932.

Kummer, Ernst: Gypsmodell der Steinerschen Flaeche. Monatsber. *Preuss. Akad.* 1863, 539.

Pinkall, Ulrich: Models of the real projective plane. In G. Fischer II, 63–67.

Rottner, Nadja and Peter Weibel: *Ruth Vollmer 1961–1978: Thinking the Line.* Ostfildern: Hatje-Cantz 2006.

Steiner, Jakob: *Gesammelte Werke II,* Nr. 32

CHAPTER 2

FROM MORPHISMS TO MATRICES

Klaus Hoechsmann
Pacific Institute for the Mathematical Sciences, Canada

Abstract: The "New Math" was an attempt to implant the spirit of Bourbaki in the classroom. In linear algebra this meant introducing coordinate-free linear mappings. However, all concrete examples involved matrices in low dimensions with few convincing applications. Forced by the dissonance between the students' expectations and his own, the author finally designed a course in matrix algebra, whose beginning and end are sketched in this article. The last section gives an elementary proof of an old theorem of Camille Jordan.

INTRODUCTION

Clashing Beliefs

In my three decades as a university professor, it frequently happened that I was assigned an introductory course in Linear Algebra. My first attempt at this, though filled to the rim with pedagogical and mathematical enthusiasm, was an unmitigated disaster. In 1967, I had just come back to Canada after ten years spent as a graduate student in the United States and post-doc in Germany, with all the innocent beliefs grown and fostered

The Montana Mathematics Enthusiast, pages 13–30

in those blessed situations. For present purposes, they can be summarized in two points: (i) Though in perpetual evolution, mathematics provides a peculiarly firm kind of knowledge which is not available in other ways; it is worth knowing "*pour l'honneur del'esprit humain*" (Dieudonné,1987), if for no other reason.(ii)The role of the mathematics teacher was to draw the student's mind into this mighty river by finding suitable entry points (exercises, examples, puzzles) and describing its sweep and flow with the greatest possible clarity. Dream on, eh? In those intervening years, I had not been insulated from teaching, but events had conspired to provide me with (small classes of) students who were mathematics enthusiasts with belief systems similar to mine. My rude awakening, which began when Günter had just turned 20, came through the encounter with larger groups of young people who marched to a very different drummer. As I grew to know them, in and after class, during office hours and at parties, they told me how they saw our subject:

(a) Though hard to understand, mathematics was an essential tool in every theoretical and practical science (hence also for academic advancement), wherefore its many facts and procedures had to be memorised, ideally forever, but in practice till the next exam.
(b) The role of the mathematics teacher was to show you, without irrelevant frills, how to handle this tool , ideally for various complex situations, but in practice for the exercises in the textbook and (especially) potential exam questions.

Such a divergence of perspective was bound to inhibit communication and had to be reduced. But how? The students were caught in a system which proclaimed Academic Freedom at convocation time, but otherwise practised paternal despotism by prescribing mathematics (among other things) and then certifying them as winners or losers via questionable "evaluation" techniques. To get into medicine in Canada, you needed Calculus (in Germany, you needed Latin), yet nobody would supply a reason I could understand. My brother-in-law, a neurosurgeon, bitterly challenged me years later to justify it. Could I have answered that it was "pour l'honneur de l'esprit humain"?

Whether or not they believed in the benefits of memorisation (which, by the way, seem to be non-trivial), the students' beliefs (a) and (b) can be seen as a rational response to the demands of the Kafkaesque machine that processed them. But where was my rationale? While there were many ways of adapting item (ii) to the realities of the day without undue corruption, I could find no honest way past the bottom line of (i): there are only two ways of knowing, namely observation and deduction, of which the second is uniquely human and most intensely cultivated in mathematics.

Since it has become fashionable to quibble with this kind of statement, I need to clarify my modest version of it, before getting back to the gap between my beliefs and the students'. Although I am strongly sympathetic to the views expressed, for instance, by Alain Connes in his famous dialogue with Jean-Pierre Changeux (Changeux and Connes, 1995), I am not defending any ideological position. I just find that there is something delightfully sturdy about well-weathered mathematics, irrespective of whether you think it is discovered or invented.

To mention one tiny example: there are exactly five regular polyhedra, having 4, 6, 8, 12, 20 faces respectively, and the dimensions of the last two are tied up with the Golden Mean. Such facts could not be changed even by the Divine Will, which is why they were so unpopular with medieval theologians. Yet they are visible to you and me in the darkest night, and any questioning of their validity only strengthens them.

Nature versus Texture

This robust nature of mathematical truth has been a constant in the over two thousand years since Archimedes (at least), and so has the basic method of pursuing it: reasoning and imagining, staring at sand-trays, black-boards, or ceilings, lost in thought. What has changed, however, is what is seen behind those ceilings. It is more than just style, as it really produces mental imagery of a different sort. For an easy example, take "Wilson's Theorem": every prime p divides the number $(p - 1)! + 1$. The present way of explaining it is to say that, in the field of p elements, the non-zero members other than 1 and -1 come in reciprocal pairs, which are knocked out in the product over all of them, leaving -1 as sole survivor. The point is that numbers now stay in the back of the mind as fleeting shadows.

In linear algebra (and analysis), many of the *dramatis personæ* of the nineteenth century, in as much as they already existed, were similarly reduced to faceless "vectors," denizens of abstract "vector spaces" tossed about by "linear mappings," a.k.a. transformations or operators, and these were later seen as special cases of "morphisms." Without this radical pruning of formerly obstructing trees, which soon gripped all of mathematics, the gigantic forests explored in the twentieth century would never even have been seen.

By the middle of the twentieth century, this recasting of mathematics, which had begun fifty years earlier, had given it a different texture or "feel," and become the dominant way of thinking about it. Through the na¨ıvet´e of most mathematicians and the wrongheaded ambitions of a few trail-blazers, the educational establishment came to believe that this way of thinking could begin in grade school. My misadventures with linear algebra (and I was not alone in this leaky boat) suggested that it was not even suitable for the college level.

Getting back to beliefs, this is not the place to describe all the little compromises in the students' position (a) and my position (ii) which made life more harmonious: they are easy to imagine and not particularly original. The hard core of disagreement was over the nature of mathematics in (i) and the *real* purpose of taking the course in (b). It gradually dawned on me that, to the students, the course content was not defined by the syllabus but by the questions on old exams. So, I went and studied them.

They had a strong family resemblence. There was always one "big" (say, 4×5) system of linear equations, whose homogeneous and inhomogeneous solutions were highly endowed with points, with a subsidiary quickie on such matters as rank and nullity, starved for points. There was also one "theoretical" question about kernel and image of a linear map, extending the basis of a vector-space and the like, also lightly weighted. The rest was about matrices and coordinatised vectors in two or three dimensions, weighted at least 3:2 in favour of the lower dimension.

Compared to Calculus, linear algebra had at least two extra hurdles for the learner (engineering students called it "Mystery Math"): it lacked a central theme, like the connection between derivatives and integrals, and it always vacillated between the parallel worlds of vector-spaces and matrices, making it difficult for mental images to jell. In reorganising the course to fit the exam, writing my own slim set of notes with exercises and all, I came down on the side of matrices and chose "standard forms" as the central theme.

Spending one third of the course on 2×2 matrices, allows students to become thoroughly familiar of working with matrix phenomena which also occur in higher dimensions. In particular, eigenvectors quickly show up in the kind of "phase-plane" diagrams you see in books on differential equations (cf. V.I. Arnold, 1980, p.26 ff). After all, when you solve the latter by Euler's Method, they turn into difference equations, whose solution orbits are of the form $\{A^n X \mid n = 0, 1, 2, \ldots\}$, where A is a matrix (in the linear case). The programming needed to make these visible on a computer screen is trivial, and playing with the program is fascinating. The question of eigenvalues comes up naturally, and is easily handled by quadratic equations.

Instead of such diagrams, this article includes two stories (*Ahorita Vengo* and *I conigli di Leonardo*) on this theme. It then shows, in a somewhat compressed manner, how the necessary algebra was delivered. Of course, everything gets more difficult with the step up to three dimensions, but the basic drift can be maintained. The final chapter, which I got to only once, with a particulary bright class, deals with standard forms for $n \times n$ matrices. An elementary rendition of the Jordan form ends the paper.

PRELIMINARIES

A *matrix* is just a rectangular array of numbers, in itself nothing mysterious. However, the algebraic interactions of such arrays, and their geometric manifestations, open up a world of fascinating mathematical phenomena. To keep things simple, we'll initially limit ourselves by and large to just two kinds of matrices:

$$A = \begin{bmatrix} a & b \\ c & d \end{bmatrix} \quad \text{and} \quad X = \begin{bmatrix} x \\ y \end{bmatrix}.$$

When we say "matrix," we shall usually mean the first of these types, preferring the term "vector" for the second.

Vector quantities abound. For instance, instead of considering an investor's assets as a single amount, you might break it down into separate values for stocks and for bonds; or, instead of looking at a total population of eagles, you might count adults and juveniles as separate subpopulations. To study how these numbers change from one time period to the next, you generally have to consider four coefficients: the growth rates of x and y by themselves, as well as the factors by which they hinder or enhance each other. This is where matrices come in; most frequently they occur as multipliers of vector quantities. The multiplication of the vector X by the matrix A shown above is defined as follows:

$$AX = \begin{bmatrix} a & b \\ c & d \end{bmatrix} \begin{bmatrix} x \\ y \end{bmatrix} = \begin{bmatrix} ax + by \\ cx + dy \end{bmatrix}.$$

If the terms b and c are both $= 0$, the matrix A is called *diagonal.* In that case, the components of AX are just ax and dy with no cross-terms, and the problem does not really need vectors; it is just one ordinary 1-dimensional problem on top of another. One of our major themes is the search for ways of simulating this happy state of affairs ("diagonalization") or at least to come as close to it as possible.

Before we can go anywhere on this meager lead, we need a minimum of information on matrix algebra. Of course, matrices can be added together and scaled in the most obvious manner (entry-by-entry), but multiplication should respect the action on columns ("vectors") given above:

$$AB = \begin{bmatrix} a & b \\ c & d \end{bmatrix} \begin{bmatrix} x & u \\ y & v \end{bmatrix} = \begin{bmatrix} ax + by & au + bv \\ cx + dy & cu + dv \end{bmatrix}.$$

If we think of B as two columns side by side, say $B = [X, V]$, multiplication by A amounts to $AB = [AX, AV]$. It is amazing that this makes 2×2 matrices into a viable algebraic system. The interplay of rows and columns is necesary for the "associative" property $A(BC) = (AB)C$, but that is not as obvious as one might wish.

On the other hand, a careful look at the definition reveals that the two distributive laws, namely, $A(B+C) = AB + AC$ and $(A+B)C = AC + BC$, do in fact hold. However, we shall presently find examples to show that $AB \neq BA$ in general, and that $AB = 0$ is possible even if $A \neq 0$ and $B \neq 0$. Such examples come up already if one of the factors is a *diagonal* matrix $D(\alpha, \beta)$:

$$\begin{bmatrix} \alpha & 0 \\ 0 & \beta \end{bmatrix}\begin{bmatrix} a & b \\ c & d \end{bmatrix} = \begin{bmatrix} \alpha a & \alpha b \\ \beta c & \beta d \end{bmatrix} \quad \text{and} \quad \begin{bmatrix} a & b \\ c & d \end{bmatrix}\begin{bmatrix} \alpha & 0 \\ 0 & \beta \end{bmatrix} = \begin{bmatrix} \alpha a & \beta b \\ \alpha c & \beta d \end{bmatrix}.$$

Note: left multiplication by a diagonal scales the rows, while right multiplication by a diagonal scales the columns of the other matrix. If $\alpha = \beta = 1$, we have the *identity matrix* $I = D(1, 1)$ which, as a multiplier, has no effect at all: $IA = AI = A$ for any A. Before doing any more symbolic work, let us look at two illustrations.

Ahorita Vengo

This is the story of a ficticious Mexican capitalist, a lady named Ahorita Vengo, who holds considerable sums of both Canadian and American dollars, but must do all her calculations and transactions in Mexican pesos. If x and y denote the number of millions of pesos invested in Canadian and American currencies, respectively, the annual growth of these monies can be represented by the matrix multiplication

$$A\begin{bmatrix} x \\ y \end{bmatrix} = \begin{bmatrix} 1.09 & -.04 \\ -.03 & 1.05 \end{bmatrix}\begin{bmatrix} x \\ y \end{bmatrix} = \begin{bmatrix} 1.09x - .04y \\ -.03x + 1.05y \end{bmatrix}.$$

For now, you may wish to skip the derivation of this equation, but if you are curious, here is how it works: firstly, there is a higher annual interest rate in Canada (12%) than in the US (9%); secondly, the Mexican government taxes Canadian holdings at a lesser yearly rate (6%) than the American ones (8%); thirdly, these taxes on North American accounts are payable half in Canadian dollars and half in US dollars, irrespective of their source.

Ahorita is annoyed by the irregularity of this growth pattern. If she puts 3 million pesos into Canadian ($x = 3$) and an equal amount into American ($y = 3$), she winds up, after one year, with 3.15 in one currency and 3.06 in the the other. Unfortunately this does not mean that x consistently grows at

5% and y at 2%. If she had split her 6 million by putting $x = 4$ and $y = 2$, the results would have been 4.26 and 1.98, respectively, suggesting growth rates of 6.5% and –1%. Not even the totals are the same: the 6 would have grown to 6.21 under the first scheme and to 6.24 under the second.

Her sister-in-law, Megusta Poco, is a professional broker and tells her about the magic 40–60 split, suggesting that she invest 2 million in Canadian and 3 million in US money. Sure enough, if she puts $x = 2$ and $y = 3$, she gets 2.06 and 3.09 for the next year, a clean 3% gain across the board. Best of all, the 2:3 ratio is maintained, and the same percentage increase will be repeated every year, great! She leaves the 5 million in Poco's care and turns her attention to personal matters.

She still has 1 million left to play with, so a few days later she goes to consult Poco's archrival Loquiero Mucho, a somewhat shady dude, feared and respected for his uncanny knack of making money. Mucho says that there is another magic ratio, and on receipt of a hefty fee divulges it as 2: –1. What does this mean? Take your million, he says, borrow another million in US funds (this makes $y = -1$), and put the 2 million into Canadian dollars (so $x = 2$). Lo and behold: Ahorita gets 2.22 for x and –1.11 for y after one year, the same 11% in both components and, of course, for the total. Again the ratio is maintained, and the same fabulous return will swell her account year after year.

Ms. Vengo's first impulse is to take all her money out of Poco's hands and give it to Mucho, but then she remembers her family ties and also the nasty law which forbids her making a net debt in a foreign currency. At least she is pleased with the neatness of the new arrangement. Instead of getting confused in the vagaries of *two currencies*, she now thinks in terms of *two portfolios*: Mucho's consistently grows at 11% and Poco's at 3%. It is as if she were working in a simpler monetary system: Mucho marks and Poco pounds.

What Happened?

Ahorita's brokers had stumbled onto the two ratios 2: –1 and 2:3 defining the "eigenlines" of the matrix A. If we take vectors V and W from each of these lines and stick them together into a new matrix $M = [V, W]$ we get

$$AM = A\begin{bmatrix} 2 & 2 \\ -1 & 3 \end{bmatrix} = \begin{bmatrix} 2 & 2 \\ -1 & 3 \end{bmatrix}\begin{bmatrix} 1.11 & 0 \\ 0 & 1.03 \end{bmatrix} = MD,$$

because A affects V like multiplication by 1.11 (Loquiero's "eigenvalue"), and likewise W by 1.03 (Megusta's). The matrix M acts as a kind of dictionary from the new coordinates X' based on portfolios to the old ones $X = MX'$ based on currencies. If X is changed to AX, what happens to X'? Well, $AMX' = M(DX')$ says that X' has changed to DX'.

This would be clearer, if we had a matrix M^{-1} which undoes multiplication by M. Then $X = MX'$ would be the same as $M^{-1}X = X'$, in other words, M^{-1} would translate from old coordinates X to new coordinates X'. Multiplying $AMX' = M(DX')$ through by M^{-1}, we'd get $M^{-1}AMX' = DX'$, which (read right to left) says:

> converting portfolio coordinates X' to currency ones MX', then applying the Mexican matrix A to obtain AMX', and finally reconverting to portfolio by applying M^{-1} is a waste of time, because it all just amounts to the easy diagonal multiplication DX'.

In short, M^{-1} changes coordinates into a system which "diagonalises" the messy opertion A. But does it exist? Sure: if we had enough matrix eperience, we would already know that

$$\begin{bmatrix} 3/8 & -1/4 \\ 1/8 & 1/4 \end{bmatrix}\begin{bmatrix} 2 & 2 \\ -1 & 3 \end{bmatrix} = \begin{bmatrix} 1 & 0 \\ 0 & 1 \end{bmatrix},$$

but it's never too late to check it. The matrix with the fractions is M^{-1}, the "inverse" of M, and the one on the right hand side is, of course, the "identity." The whole equation is written as $M^{-1}M = I$. And yes: multiplying $AM = MD$ by M^{-1} formally yields $M^{-1}AM = M^{-1}MD = ID$, but the I quickly disappears, since multiplying by it changes nothing.

I conigli di Leonardo

In many biological species, population growth can be studied by keeping track of two types of females: *adults* and *juveniles*. Juveniles are those who were born in the last breeding season and survived to the present one. Presumably they produce less offspring than adults (often none at all) and survive less frequently to the next breeding season (in which they will be adults). Males are ignored in this kind of study, apart from being assumed available to fill out the reproductive capacities of the females.

Let x_n be the the number of adults, y_n the number of juveniles, and $t_n = x_n + y_n$ the total number of females (excluding babies) in the n-th season. What will be the situation in the following season? If a and b denote the survival rates for adults and juveniles, respectively, the number of adults will be $x_{n+1} = ax_n + by_n$. If every adult contributes (on the average) c juveniles to the next season, and every juvenile likewise contributes d future juveniles (usually $d = 0$), the younger set will number $y_{n+1} = cx_n + dy_n$. This can be summarized in the matrix equation

$$\begin{bmatrix} x_{n+1} \\ y_{n+1} \end{bmatrix} = \begin{bmatrix} a & b \\ c & d \end{bmatrix} \begin{bmatrix} x_n \\ y_n \end{bmatrix}.$$

Abbreviating the square matrix by A and the n-th "population vector" by X_n, this simplifies to $X_{n+1} = AX_n$, which could be jazzed up to $AX_n = A^2X_{n-1} = \cdots = A_nX_0$,

For any A with $a, b, c > 0$, and $d \geq 0$, the evolution of a non-negative population vector $X_n = A^nX_0$ happens to be governed by two facts:

1. As n increases, the fraction x_n/t_n approaches a constant x^* independent of the initial x_0, y_0. Of course, y_n/t_n will simultaneously approach $y^* = 1 - x^*$. We call this the *stable distribution* and any X with $x/t = x^*$ and $y/t = y^*$, a *stable vector.*
2. If X is a stable vector then so is AX. This means that, on a stable vector, multiplication by A has a very simple effect. Since the ratio between the components cannot change, both of them get multiplied by the same *growth factor* $\lambda > 0$. So, a stable X "experiences" A as if it were a simple number instead of a matrix: $AX = \lambda X$.

Taken together, these tell us that, if we wait long enough, X_n will approach stability and stay there. From then on our problem reduces to the familiar one-dimensional growth (or decay) of $t_n = x_n + y_n$ by the factor $\lambda > 0$. Our efforts must therefore be directed toward finding the stable distribution and the growth factor. Actually, we'll do this in reverse: first find λ and then solve the equation $(A - \lambda I)X^* = 0$ subject to the condition $x^* + y^* = 1$.

Example 1

For some species of birds, the reproduction story outlined above might yield a matrix with $a = .7$, $b = .3$, $c = 2$, $d = 0$. Repeated squaring yields

$$A^2 = \begin{bmatrix} 1.09 & 0.21 \\ 1.40 & 0.60 \end{bmatrix} \quad A^4 = \begin{bmatrix} 1.48 & 0.35 \\ 2.37 & 0.65 \end{bmatrix} \quad A^8 = \begin{bmatrix} 3.04 & 0.76 \\ 5.05 & 1.27 \end{bmatrix}.$$

From the last power we see that the x-part of both columns seems to go toward $3/8 = .375$. This is borne out by checking A^{16}, the square of A^8. Thus $x = 3$, $y = 5$ should give a stable vector. Indeed multiplication by A yields $(.7)(3) + (.3)(5) = 3.6$ and $(2)(3) + (0)(5) = 6$, which is in the same ratio 3:5 but multiplied by $\lambda = 1.2$.

Example 2

One of the earliest "modern" European mathematicians (ca. 1200 AD) was Leonardo da Pisa, alias Fibonacci, the man who introduced Indo-Arabic numerals to the west. Today he is remembered mainly for the numerical sequence

$$1, 1, 2, 3, 5, 8, 13, 21, 34, 55, 89, \ldots,$$

generated by the powers of the matrix A which has $a = b = c = 1$, and $d = 0$. It is remarkable that Leonardo actually posed his original problem in the context of the reproduction of rabbits (conigli in Italian), immortal ones with a survival rate = 1. Here we have

$$A^4 = \begin{bmatrix} 5 & 3 \\ 3 & 2 \end{bmatrix} \quad A^8 = \begin{bmatrix} 34 & 21 \\ 21 & 13 \end{bmatrix} \quad A^{16} = \begin{bmatrix} 1597 & 987 \\ 987 & 610 \end{bmatrix}.$$

In the 16-th power the ratios for the first and second columns are both equal to 1.61803. This also happens to be the growth factor, whose theoretical value is $\lambda = (1+\sqrt{5})/2$.

A Long Leap Forward

The most obvious method for finding λ is to look at increasing powers of A until we see the same ratio jelling in both columns. To get high powers of A quickly, it is advisable to square over and over, forming $A^2 = AA$, $A^4 = (A^2)^2$, $A^8 = (A^4)^2$, $A^{16} = (A^8)^2$, and so on, as we have done above.

We'll soon discover a way of directly computing the growth factor (a.k.a. "eigenvalue") as

$$\lambda = \frac{1}{2}\left[(a+d) + \sqrt{(a-d)^2 + 4bc}\,\right]. \tag{2.1}$$

by solving a quadratic equation. Of course, such equations usually have two solutions, the second one being obtained by replacing the + in front of the root by a minus sign. We would get -0.5 in the first example and $(1-\sqrt{5})/2$ in the second, neither of which would make much sense in terms of the model. However, they do allow a formal diagonalisation, which we shall now explore for the case of Example 2.

Putting $(1-\sqrt{5})/2 = \mu$, following the prescriptions for finding eigenlines, and forming M, we get

$$AM = \begin{bmatrix} 1 & 1 \\ 1 & 0 \end{bmatrix}\begin{bmatrix} \lambda & \mu \\ 1 & 1 \end{bmatrix} = \begin{bmatrix} \lambda & \mu \\ 1 & 1 \end{bmatrix}\begin{bmatrix} \lambda & 0 \\ 0 & \mu \end{bmatrix} = MD,$$

which is so slick and tidy because $\lambda + \mu = 1$ and $\lambda\mu = -1$. Together with $\lambda - \mu = \sqrt{5}$, these relations also allow us to verify

$$\frac{1}{\sqrt{5}}\begin{bmatrix} 1 & -\mu \\ -1 & \lambda \end{bmatrix}\begin{bmatrix} \lambda & \mu \\ 1 & 1 \end{bmatrix} = \begin{bmatrix} 1 & 0 \\ 0 & 1 \end{bmatrix},$$

which shows that the first of these matrix factors (*with* $\sqrt{5}$ in the denominator) is M^{-1}. The point of all this is to demonstrate another use of diagonalisation: computing powers. Note that $(M^{-1}AM)^2 = M^{-1}AMM^{-1}AM = M^{-1}A^2M$ because the $M^{-1}M$ in the middle wipes out. And so it goes for *all* powers: $(M^{-1}AM)^n = M^{-1}A^nM = D^n$.

To know the $(n + 1)$*st number in the Fibonacci sequence*, for instance, you can figure out that it will be top left entry of A^n. You then use your pocket calculator to find D^n and translate back by $A^n = MD^nM^{-1}$, like so:

$$\begin{bmatrix} 1 & 1 \\ 1 & 0 \end{bmatrix}^n = \begin{bmatrix} \lambda & \mu \\ 1 & 1 \end{bmatrix}\begin{bmatrix} \lambda^n & 0 \\ 0 & \mu^n \end{bmatrix}\begin{bmatrix} 1 & -\mu \\ -1 & \lambda \end{bmatrix} \cdot \frac{1}{\sqrt{5}}.$$

For the first entry of the result, you'll get $(\lambda^{n+1} - \mu^{n+1}) / \sqrt{5}$. Since this has to be an integer and since $|\mu^n|$ is small for sizable n (for $n = 16$, it is less than 0.0005), the rule is this: just compute $\lambda^{n+1} / \sqrt{5}$ and take the nearest integer. For $n = 16$, the calculator says 1596.99987, which agrees nicely with the 1597 we got in Example 2 above.

MORE ALGEBRA

The time has come to stop the hand-waving and get a little more precise. We shall do the minimum to give clean explanations of what happened in Section 2 and motivate what we are about to do in Section 4.

Inverses

The easy equation

$$\begin{bmatrix} a & b \\ c & d \end{bmatrix}\begin{bmatrix} d & -b \\ -c & a \end{bmatrix} = \begin{bmatrix} ad - bc & 0 \\ 0 & ad - bc \end{bmatrix} = (ad - bc)I \qquad (2.2)$$

turns out to be a gold mine of information. The second factor appearing in it is visibly concocted from the first factor A, and is known as its *adjoint*, denoted A^*. The scalar $(ad - bc)$ on the right hand side is called the *determinant*, det A. Equation (2.2) can thus be abbreviated as $AA^* = (\det A)I$, and one easily checks that $A^*A = (\det A)I$, as well.

If $\det A \neq 0$, the matrix A has an *inverse*

$$A^{-1} = \frac{1}{\det A}A^* \quad \text{with} \quad AA^{-1} = A^{-1}A = I.$$

If A has an inverse, any equation $AX = 0$ clearly implies $X = 0$. On the other hand, if A does not have an inverse, then $AA^* = 0$, and $AX = 0$ when X is either column of A^*. Hence we can say that

A has an inverse if and only if $AX = 0$ is impossible with $X \neq 0$.

It should also be noted that inversion reverses the order of factors, i.e., $(AB)^{-1} = B^{-1}A^{-1}$, because $ABB^{-1}A^{-1} = AIA^{-1} = I$.

Rewriting $AA^* = (ad - bc)I$ with $A^* = (a + d)I - A$ yields the *Cayley-Hamilton relation*

$$A^2 - (a + d)A + (ad - bc)I = 0. \tag{2.3}$$

The scalar $(a + d)$ is called the *trace* of A and denoted by tr A.

Computing the determinant of the product AB, we obtain the miraculous result $(ax + by)(cu + dv) - (cx + dy)(au + bv) = (ad - bc)(xv - yu)$, i.e.,:

$$\det(AB) = (\det A)(\det B). \tag{2.4}$$

Eigenstuff

A 2-column $V \neq 0$ is an *eigenvector* of A if multiplication by A only scales it by some numerical value λ, i.e., $AV = \lambda V$ or $(A - \lambda I)V = 0$ An eigenvector never comes alone. If $AV = \lambda V$ then also $AW = \lambda W$ for every scalar multiple $W = \alpha V$. Hence the whole "eigenline" determined by V experiences the action of A as a simple scaling by the eigenvalue λ.

There are several ways of computing eigenvalues. For 2×2 matrices they are easily obtained by applying the Cayley-Hamilton relation (2.3) to a putative eigenvector V and obtaining $(\lambda 2 - (\text{tr } A)\lambda + \det A)V = 0$. Since V must be $\neq 0$ to qualify, must satisfy the *characteristic* equation $\lambda^2 - (\text{tr } A)\lambda + \det A = 0$, which simplifies to

$$(\lambda - r)^2 = q \quad \text{with} \quad \begin{aligned} 2r &= a + d, \\ 4q &= (a - d)^2 + 4bc, \end{aligned} \tag{2.5}$$

and has at most two solutions, namely $r \pm \sqrt{q}$. This explains Equation (2.1). Once these λ are determined, it is easy to solve $(A - \lambda I)V = 0$ for the actual eigenvectors V. Note: tr A is the sum, and det A is the product of $\lambda_1 = r + \sqrt{q}$ and $\lambda_2 = r - \sqrt{q}$.

The extreme case $A - \lambda I = 0$ needs no further clarification and will be excluded from the following discussion. Otherwise, three cases can occur:

(a) There might be two different eigenlines, generated by (say) V and W. The matrix $M = [V, W]$ is then invertible, because $MX = 0$ is impossible (with $X \neq 0$), as it would be of the form $aV + bW = 0$. Then we get a diagonalisation $M^{-1}AM = D$, as in the illustrations of the preceding section. This happens for $q > 0$.
(b) The eigenvectors of A form a single line. Of course, the effect of A then must be the same on every eigenvector. Therefore $\lambda_1 = \lambda_2$, and $q = 0$.
(c) There are no (real) eigenvalues and eigenvectors, i.e., $q < 0$.

If we are allowed to use *complex* numbers, even $q < 0$ yields two eigenvalues and hence a diagonalisation $A = MDM^{-1}$ with complex matrices M and D.

Similarity

The matrices A and B are said to be similar if there is an invertible matrix M such that

$$B = M^{-1}AM \quad \text{or} \quad A = MBM^{-1}.$$

It is easy to see that, if two matrices are similar to a third, they are similar to each other.

What about eigengadgets? If A and B are as above, $AMV = MBV$ readily shows that V is an eigenvector for B if and only if MV is one for A, with the *same* eigenvalue.

In particular, *similar matrices have the same trace and determinant.* If you wish to avoid complex numbers completely, you can deduce this by applying M and M^{-1} to the Cayley-Hamilton relation, and see that it does not change. Hence similar matices will fall into the same one of the patterns (a), (b), or (c) outlined above.

Given any A, we shall now show a way of constructing an invertible matrix M such that the resulting $B = M^{-1}AM$ takes on one of the following three standard forms:

$$\text{(a)} \begin{bmatrix} \lambda_1 & 0 \\ 0 & \lambda_2 \end{bmatrix} \quad \text{(b)} \begin{bmatrix} r & 1 \\ 0 & r \end{bmatrix} \quad \text{(c)} \begin{bmatrix} r & -s \\ s & r \end{bmatrix}. \qquad (2.6)$$

In the diagonalizable case (a), we already know what to do: take the columns of M to be eigenvectors of A. In the other two cases we shall exploit the Cayley-Hamilton relation, written in the form

$$(A - rI)^2 = qI, \tag{2.7}$$

with r and q as in (2.5). If there is just one single eigenline ($q = 0$ and $r = \lambda$), choose $W \neq 0$ to lie outside it, so that $V = (A - rI)W \neq 0$. Then $(A - rI)V = 0$ because of (2.7), and V is an eigenvector. Putting $M = [V, W]$, we have $(A - rI)M = [0, V] = MJ_2$, where J_2 is the 2×2 matrix with 1 in the upper right and 0 elsewhere. J_r is exactly what you get by subtracting rI from the matrix in the middle of (2.6). The invertibility of M is ensured by the fact that its columns are not aligned, making $MX = 0$ impossible (for $X \neq 0$) and $\det M \neq 0$ (cf. the paragraph on inverses).

The case (c) really belongs under the heading of commplex diagonalisation, and is included here in its present form only for the sake of completeness. Grab any vector $Y \neq 0$ (there *are* no eigenlines to stay away from), and put $M = [V, W]$ with $W = sY$, where $s^2 = -q$, and $V = (A - rI)Y$. Then $(A - rI)M = [qY, sV] = s[-W, V]$. The rest is exercise.

THE JORDAN DECOMPOSITION

Over any field of scalars which contains all roots of the minimal (or characteristic) polynomial of an $n \times n$ matrix A, it is fairly straightforward to find an upper triangular matrix B similar to A. The final step, valid over *any* field, is to go from there to the "Jordan Form."

Jordan Blocks

For every integer $r > 0$, let J_r denote the $r \times r$-matrix obtained by augmenting the $(r - 1)$-st identity matrix I_{r-1} by a trivial first column and last row (and taking $J_1 = 0$). Any matrix of the form $\lambda I_s + J_s$, where λ is a scalar, will be called the *Jordan block of degree s and proper value* λ. As an example, consider

$$\lambda I_4 + J_4 = \begin{bmatrix} \lambda & 1 & 0 & 0 \\ 0 & \lambda & 1 & 0 \\ 0 & 0 & \lambda & 1 \\ 0 & 0 & 0 & \lambda \end{bmatrix}. \tag{2.8}$$

Up to similarity, such matrices turn out to be the building blocks for *all* upper triangular matrices, in the following sense.

Theorem: *Any upper triangular $n \times n$ matrix A is similar to a "direct sum" of Jordan blocks $A_0, \ldots, A_l$, meaning that*

$$M^{-1}AM = \begin{bmatrix} A_0 & 0 & 0 \\ 0 & \ddots & 0 \\ 0 & 0 & A_l \end{bmatrix} \tag{2.9}$$

for suitable $n \times n$ matrix M.

Prelude

Before jumping into the belly of the proof, let us clear the deck of routine material.

1. *Preparing an induction.* Since M would commute with any cI_n, replacing A by $A - a_{11}I_n$ does not really change the problem, and we may therefore assume that the first column of A is zero. To the right of this column and below the first row of A, we see an upper triangular $(n-1) \times (n-1)$ matrix B, which by induction we may suppose to have already transfomed into a direct sum of Jordan blocks $B_1, \ldots, B_m$. Altogether we have

$$\begin{bmatrix} 1 & 0 \\ 0 & N^{-1} \end{bmatrix} A \begin{bmatrix} 1 & 0 \\ 0 & N \end{bmatrix} = \begin{bmatrix} 0 & R_1 & R_2 & \cdots & R_m \\ 0 & B_1 & 0 & \cdots & 0 \\ 0 & 0 & B_2 & \cdots & 0 \\ \vdots & \vdots & \vdots & \ddots & \vdots \\ 0 & 0 & 0 & \cdots & B_m \end{bmatrix} = A', \tag{2.10}$$

where N is the $(n-1) \times (n-1)$ matrix which had brought B into the desired form. At this point, the first column of our A' matrix is still 0, and the first row has the form $[0, R_1, \ldots, R_m]$, with the subrow R_k sitting above the block B_k. We shall need to remove these "obstructions" R_k.

2. *Permuting the blocks.* Whenever we permute the B_k by suitable similarities of A', the corresponding R_k are permuted accordingly. This follows from the fact that permutations are just reshuffling the index set with which rows and columns are labelled. As the index 1 is not affected by the shuffle, the first column stays as it is (namely zero), whereas the first row is rearranged. If that sounds too vague, you may wish to verify the following equation:

$$\begin{bmatrix} I_r & 0 & 0 & 0 \\ 0 & 0 & I_t & 0 \\ 0 & I_u & 0 & 0 \\ 0 & 0 & 0 & I_s \end{bmatrix} \begin{bmatrix} * & X & Y & * \\ 0 & A & 0 & 0 \\ 0 & 0 & B & 0 \\ 0 & 0 & 0 & * \end{bmatrix} \begin{bmatrix} I_r & 0 & 0 & 0 \\ 0 & 0 & I_u & 0 \\ 0 & I_t & 0 & 0 \\ 0 & 0 & 0 & I_s \end{bmatrix} = \begin{bmatrix} * & Y & X & * \\ 0 & B & 0 & 0 \\ 0 & 0 & A & 0 \\ 0 & 0 & 0 & * \end{bmatrix}. \quad (2.11)$$

Since every permutation can be achieved by systemeatically switching adjacent elements (think of books on a shelf), this apparently special case establishes our claim.

The Key Lemma

Left multiplication by J_r will push the entries of a column upward, losing the first and replacing the last by 0, while right multiplication by it will push the entries of a row to the right, losing the last and replacing the first by 0. These properties combine to yield the following magic, with $b + b' = c + c' = 0$ for typographical reasons:

$$\begin{bmatrix} 0 & 1 & 0 & 0 \\ 0 & 0 & 1 & 0 \\ 0 & 0 & 0 & 1 \\ 0 & 0 & 0 & 0 \end{bmatrix} \begin{bmatrix} b' & c' & 0 \\ a & 0 & 0 \\ 0 & a & 0 \\ 0 & 0 & a \end{bmatrix} - \begin{bmatrix} b' & c' & 0 \\ a & 0 & 0 \\ 0 & a & 0 \\ 0 & 0 & a \end{bmatrix} \begin{bmatrix} 0 & 1 & 0 \\ 0 & 0 & 1 \\ 0 & 0 & 0 \end{bmatrix} = \begin{bmatrix} a & b & c \\ 0 & 0 & 0 \\ 0 & 0 & 0 \\ 0 & 0 & 0 \end{bmatrix}. \quad (2.12)$$

LEMMA: *Let $r > s$, and consider any $r \times s$ matrix Z whose non-zero entries are all in the first row. Then there exists an $r \times s$ matrix X such that $J_r X - X J_s = Z$.*

Proof. Equation (2.12) shows how X can be constructed: if Z has $[a_1\, a_2 \ldots a_s]$ as its first row, we use a_1 for the "subdiagonal" entries of X and $[-a_2 \ldots -a_s\, 0]$ for the first row.

Proof of Theorem

The main tool for "killing" the obstructions R_k is the easily verified formula

$$\begin{bmatrix} I_t & -X \\ 0 & I_u \end{bmatrix} \begin{bmatrix} T & Y \\ 0 & U \end{bmatrix} \begin{bmatrix} I_t & X \\ 0 & I_u \end{bmatrix} = \begin{bmatrix} T & Z \\ 0 & U \end{bmatrix} \quad \text{with} \quad Z = Y + TX - XU, \quad (2.13)$$

which is valid for square matrices T and U of degrees t and u, respectively, and $t \times u$ matrices X, Y, Z. Taking the middle factor on the left to be the identity matrix yields $Z = 0$, and shows that this is indeed a similarity relation. It will be used in both steps of the proof.

Step 1

To begin with, we only reduce the obstructions in the most obvious way. In (2.13), we take $U = B_m$, whence Y is zero below the first row (which is $= R_m$). Then T must be the $(n-u) \times (n-u)$ submatrix of A' lying "north-west" of B_m. Setting the last $t-1$ rows of X to zero we get $TX = 0$ because of the zeroes in the first column of T. Thus we obtain $Z = Y - XU$, which really says $Z_* = R_m - X_*B_m$, where X_* and Z_* denote first rows, the remaining ones being zero.

If B_m is invertible, Z_* can clearly be made zero by a suitable X_*. If *not*, B_m must be $= J_u$, and X_*B_m can be made into any u-tuple with 0 as first coordinate. Hence Z_* can always be reduced to having at most one non-zero entry, namely, the first one. If $Z_* = 0$ for any reason, our game is won by induction, because we can consider the puzzle solved for the smaller matrix T. Since we are allowed to switch B_m with any B_k, we can apply the same argument to all of them.

Step 2

If the game is not yet over, we are faced with a situation which still looks like (2.10), but where every R_k is of the form $[c_k, 0, \dots, 0]$ with $c_k \neq 0$. Moreover, each $B_k = J_{d(k)}$ is a Jordan block with proper value 0. Let us arrange them in decreasing size, and make sure that $R_1 = [1, 0, \dots, 0]$, by multiplying the first row of A' by $1/c_1$ (and its trivial first column by c_1, leaving it unaffected but producing a correct similarity). Now comes the surprise: the upper left corner of A′ looks like

$$\begin{bmatrix} 0 & R_1 \\ 0 & J_{d(1)} \end{bmatrix} = \begin{bmatrix} 0 & 1 & 0 & \cdots & 0 \\ 0 & 0 & 1 & \cdots & 0 \\ \vdots & \vdots & \vdots & \ddots & \vdots \\ 0 & 0 & 0 & \cdots & 1 \\ 0 & 0 & 0 & \cdots & 0 \end{bmatrix} = J_r, \tag{2.14}$$

which is none other than $J_{d(1)+1}$. The matrix displayed in (2.10) has thereby morphed into

$$A'' = \begin{bmatrix} J_r & R_2 & \cdots & R_m \\ 0 & B_2 & \cdots & 0 \\ \vdots & \vdots & \ddots & \vdots \\ 0 & 0 & \cdots & B_m \end{bmatrix}, \tag{2.15}$$

where $B_k = J_{d(k)}$ for $k > 1$, as before. With $r = d(1) + 1$ strictly greater than the other degrees $d(k)$, we can unleash the key lemma via the following expanded version of (2.13):

$$\begin{bmatrix} I_s & 0 & -X \\ 0 & I_t & 0 \\ 0 & 0 & I_u \end{bmatrix} \begin{bmatrix} S & R & Y \\ 0 & T & 0 \\ 0 & 0 & U \end{bmatrix} \begin{bmatrix} I_s & 0 & X \\ 0 & I_t & 0 \\ 0 & 0 & I_u \end{bmatrix} = \begin{bmatrix} S & R & Z \\ 0 & T & 0 \\ 0 & 0 & U \end{bmatrix}, \qquad (2.16)$$

where $Z = Y + SX - XU$. For $S = J_r$ and $U = J_{d(m)}$, the lemma shows that $Y = R_m$ can be annihilated by a suitable choice of X.

REFERENCES

Arnold, V.I. (1980) *Ordinary Differential Equations*, The MIT Press, Cambridge, Massachusetts.

Changeux, J.-P. and Connes, A. (1995) *Conversations on Mind, Matter, and Mathematics*, Princeton University Press, Princeton, N.J.

Dieudonné, J. (1987) *Pour l'honneur de l'esprit humain*, Hacchette, Paris.

CHAPTER 3

A THEORY OF TEACHING AND ITS APPLICATIONS

Alan H. Schoenfeld
University of California, Berkeley

In commemoration of Günter Törner's 60th birthday

Abstract: Research methods and perspectives from a range of disciplines including education, psychology, artificial intelligence, and economics provide a set of tools for modeling complex human decision-making in fields such as teaching. This paper describes the foundations of such research and the underpinnings of a model of teaching-in-context that explains, on a line by line basis, the decision making by teachers during hour-long classroom lessons. The existence of such models provides tools for examining and improving teaching, as well as the possibility that the decision making in other professions can be comparably modeled.

INTRODUCTION

Teaching, like many other professions—to give two very different examples, medical practice and automobile mechanics—depends on a large skill and knowledge base [1,2]. Like such fields, its practice involves a signifi-

The Montana Mathematics Enthusiast, pages 31–37

cant amount of routine activity punctuated by occasional and at times unplanned but critically important decision making—decision making that can determine the success or failure of the effort. Moreover, how one goes about work in these professions is shaped in important ways by one's conscious or unconscious beliefs and values [3,4].

Methods and perspectives from a range of disciplines, including earlier studies of knowledge organization [5], problem solving [6] and decision making [7,8] have now been combined in ways that make it possible to provide theoretical characterizations of the behavior and decision making of practitioners in fields such as teaching. The theoretical approach has been tested by the construction of models that closely match the teachers' behavior over extended episodes of performance such as hour-long lessons [9,10]. The existence of such models offers promise of improving the profession, both from the perspective of understanding accomplished performance and identifying appropriate points of intervention as teacher develop during their careers.

KNOWLEDGE ORGANIZATION

It has long been understood that human memory is associative, and that knowledge comes in associative "packages" that, depending on the field and intellectual tradition, have been called scripts [11], frames [12], routines [13], or schemata [14]. For example, early research in text comprehension showed that when people read stories, they add to the text additional information based on stereotypical experiences: for example a person reading a story about a customer in a restaurant who enjoyed a meal but left a small tip will infer that the service was poor, even though no mention was made of the service [11]; people reading the first three words of a problem statement, "A river steamer," conjectured that the rest of the problem statement would be concerned with the boat's speed moving with and against the current and the total amount of time it took to make a round trip [14]. More generally, people invoke typical scenarios, and have "default expectations" regarding what is likely to take place, even though such information is not supplied. Similarly, teachers, doctors, and automobile mechanics make extensive informal use of diagnostic information to decide on next steps in professional interactions—as do we all, in social interactions, picking up subtle details and using them, whether correctly or not, to "fill in the gaps."

PROBLEM SOLVING

By the 1990s, researchers in mathematical and scientific problem solving, and more broadly in fields where actions are clearly "goal oriented" such as writing, had come to consensus about the categories of behavior that need to be examined in order to explain why people succeed or fail in their attempts at problem solving. The literature indicated that in any problem solving attempt, some or all of the following will be essential determinants of success or failure, and that (at this broad grain size) the list of factors was complete:

> *The knowledge base.* Clearly, what you know and how you access it (cf. *knowledge organization,* above) is a major factor in how well you succeed at what you are trying to do. The knowledge base includes concepts and procedures, but also habits of mind and patterns of productive (or unproductive) behavior. But that can't be all there is, as some people with less knowledge than others manage to solve problems that the people with greater knowledge don't manage to solve.
> *Problem solving strategies.* Such strategies, first described in mathematics by George PÛlya [15] as "heuristic" strategies, are rules of thumb for making progress in understanding a problem or finding productive approaches to its solution. Heuristic strategies are seen in all fields of endeavor, and lie at the heart of many problem solving programs in artificial intelligence [16].
> *Metacognition (specifically, monitoring and self-regulation).* Effective problem solvers keep tabs on how well their attempts are going and make adjustments on the basis of their assessments. Less effective problem solvers often fail to solve problems not because they don't "know" the relevant material, but because they persevere at initial (often ill-conceived) plans, not giving themselves the opportunity to recover and pursue more profitable directions. This is one aspect of decision making (see below).
> *Beliefs, values, and orientations.* In all fields of endeavor, our conscious or unconscious biases shape what we notice and what we choose to do. As Groopman notes in [3], such often unconscious biases shape the decisions made by doctors, ruling diagnostic possibilities in or out of contention. Similarly, problem solvers may choose or abandon paths based on aspects of their prior experience [17].

What the research on problem solving offered was a *framework* for characterizing the success or failure of decision making during problem solving. What it lacked was a complete description of mechanism, explaining how and why people made the specific decisions they made. (A partial answer was provided by research on routine knowledge access, but such work failed to account for decision making when circumstances were not familiar or when things were not going well.)

DECISION MAKING

Economists, among others, have long noted that human decision making is not rational [7,8]. If it were, lotteries would go out of business; a straightforward calculation shows that the expected value of a lottery ticket is negative. The concept of *subjective expected utility* [7] provides a way of characterizing individual decision making in such cases. If the actual cost of a ticket is subjectively diminished for an individual ("this is pin money") and the value of the reward is subjectively enlarged ("I'll be rich, and I'll be able to retire"), then the subjective expected value of a lottery ticket,

$$SEV = P(loss) \times (Subjective\ cost\ of\ ticket) + P(win) \times (Subjective\ value\ of\ win),$$

is positive, and the decision to purchase a ticket makes (subjective) sense. This notion can be generalized to provide subjective evaluations of various options an individual might consider.

SYNTHESIS AND IMPLICATIONS

First, teaching and problem solving can be seen as case examples of *goal-oriented* behavior, in which an individual establishes one or more goals and then sets about trying to achieve them. This kind of goal-oriented behavior characterizes many professions. Daily activity in such professions typically consists of the routine application of skills, punctuated by occasions that call for consequential decision making. In teaching, for example, many actions are developed as standard routines: lecturing, collecting homework, having students present work at the board or engage in group work, etc. Of course, which routines are established and how they are run will be a function of the teacher's beliefs and values: What (for this teacher) is the right balance of exposition and exploration? Can these students handle being confused, or is one better off laying things out for them? Is a quiet class a sign of something good or an indication of flagging interest?

A significant proportion of teachers' classroom activities can be characterized using the literatures on knowledge organization and problem solving. Established teachers have easy access to a range of routines for implementing everyday practices such as homework and board work. In addition, less frequent but important classroom teaching events often fit the same pattern. For example, every experienced algebra teacher knows that at some point in the course, a student will say or write the incorrect expression $(a + b)^2 = a^2 + b^2$. There are many possible ways to deal with this misconception. Given a teacher's experience and beliefs, he or she may have a few "favorites" among these approaches readily accessible in memory, and circumstances may determine which of these is actually chosen. Thus routine access to knowledge and strategies, along with appropriate metacognitive monitoring, suffice both for everyday and for less frequent but important classroom situations.

However, students sometimes come up with something new, e.g., an unexpected suggestion that, if pursued, would require a significant deviation from the lesson. The various options that come to a teacher's mind, ranging from "I'll talk to you about that after class" to "let's pursue that and see where it goes" all have costs and benefits. Different teachers will assess those costs and benefits differently. However, any particular teacher's judgments can be modeled by taking into account that teacher's subjective assessments of the costs and benefits of each option. This provides a quantitative mechanism for modeling the teacher's decision's decision making in these more rare but highly consequential situations.

A combination of the "access to familiar routines and strategies" perspective from research on knowledge organization and problem solving, and the "subjective decision making at critical times" perspective from the research on decision making provides the basis for the detailed modeling of teachers' classroom actions—and it does so at the level of explaining every utterance over the course of an hour's lesson. In simplest terms, the models work as follows:

> A teacher enters the classroom with a particular body of knowledge, goals, and orientations (beliefs, values, etc.). The individual orients to the situation. Certain knowledge and routines become salient and are activated and/ or triggered, and directions are established.
>
> - Goals are established (or reinforced if they pre-existed).
> - Decisions are made, consciously or unconsciously, in pursuit of top-level goals.
> - Knowledge is selected for implementation. (Such knowledge may be in the form of scripts, routines, or schemata.) Implementation begins.
> - Monitoring (whether effective or not) takes place on an ongoing basis.
>
> The process iterates:
>
> - If a subgoal is achieved, new subgoals kick in through the routine (or script, or...).
> - If a goal is achieved, new goals kick in via decision making.
> - If the processes is interrupted or things don't seem to be going well, decision making kicks into action once again.

The modeling approach described immediately above has been used to describe, in fine detail, a traditional high school lesson conducted by a student teacher [18]; an innovative high school lesson on measures of central tendency [19,20]; a college problem solving course, and a third grade discussion of the properties of even and odd numbers [21,22]. Despite the variations in teacher style and experience, in students' age, and in classroom topics, teachers' in-the-moment decision making in all of these classrooms was captured in detail via the same basic approach, combining a dependence on (personally selected) routines with a form of expected value computation where the outcomes are assigned values consistent with the teachers' subjective views. In short, teachers' in-the-moment decision making (and, hypothetically, decision making in other professions with a similar balance of routine and consequential non-standard decisions) can be modeled.

The ability to capture individual teachers' decision making through such models has a range of potential applications. First, an understanding of the importance of various factors shaping such decisions, e.g., teachers' beliefs, has the potential for productive interventions: as in the case of problem solving [6], teachers' classroom actions are more malleable when the teachers become more aware of the (sometimes previously unconscious) causes of those actions and more reflective about them [23]. Although the modeling process itself (at the level of line-by-line analyses) is very time consuming, its uses need not be: the same analytic approach can be used for discussions between teachers and their professional development coaches regarding which aspects of the teachers' practices are working effectively, and which could use some work. Long-term, this kind of modeling can be used to describe and analyze typical "developmental trajectories" for teachers as they become more experienced and accomplished, and to identify points in their careers when the teachers might be particularly receptive to (or need) particular kinds of interventions in order to keep developing as professionals. More broadly, given that the combination of knowledge-based and in-the-moment decision making in teaching is similar to that in a range of other professions, there is the possibility that the approach described here can be worked out in those fields as well.

REFERENCES

1. L. Darling-Hammond & J. Bransford, Eds., *Preparing teachers for a changing world.* (San Francisco, Jossey-Bass, 2005).
2. M. Chi, R. Glaser, & M. Farr, Eds., *The nature of expertise.* (Mahwah, NJ, Erlbaum, 1988).
3. J. Groopman, *How doctors think.* (New York, Houghton Mifflin, 2007).

4. M. Lampert, *Teaching Problems and the Problem of Teaching.* (New Haven, Yale University Press, 2001).
5. D. Rumelhart & A. Ortony, The representation of knowledge in memory, in *Schooling and the acquisition of knowledge,* R. Anderson, R. Spiro, & W Montague, Eds. (Mahwah, NJ, Erlbaum, 1977), pp. 99–135.
6. A. Schoenfeld, *Mathematical problem solving.* (Orlando, FL, Academic Press, 1985).
7. L. Savage. *The Foundations of Statistics.* (New York, Wiley, 1954).
8. D. Kahneman, P. Slovic, & A. Tversky. Eds., *Judgment under Uncertainty: Heuristics and Biases.* (Cambridge, Cambridge University Press, 1982).
9. *Examining the Complexity of Teaching.* Special issue of the *Journal of Mathematical Behavior,* **18**, 3 (1999).
10. *Issues in Education,* **4**, 1 (1998).
11. D. Hinsley, J. Hayes, & H. Simon, From words to equations: meaning and representation in algebra word problems. In *Cognitive processes in comprehension,* M. Just & P. Carpenter, Eds. (Hillsdale, NJ: Erlbaum, 1977), pp. 89–106.
12. Minsky, M. A framework for representing knowledge, in *The psychology of computer vision,* P. Winston, Ed. (New York: McGraw-Hill, 1977), pp. 170–195.
13. G. Leinhardt. On teaching, in *Advances in instructional psychology* (Vol. 4,), R. Glaser, Ed. (Hillsdale, NJ: Erlbaum, 1993), pp. 1–54.
14. R. Schank, & R. Abelson, *Scripts, plans, goals, and understanding.* (Hillsdale, NJ, Erlbaum, 1977).
15. G. Polya, *How to solve it.* (Princeton, Princeton University Press, 1945; 2nd edition, 1957).
16. R. Groner, M. Groner, M., & W. Bischof, Eds. *Methods of heuristics* (Hillsdale, NJ, Erlbaum, 1983).
17. Lampert, M. When the problem is not the problem and the solution is not the answer: Mathematical knowing and teaching. *American Educational Research Journal, 27, 1,* pp. 29–63 (1990).
18. D. Zimmerlin & M. Nelson. The detailed analysis of a beginning teacher carrying out a traditional lesson. *Journal of Mathematical Behavior,* **18**, 3, 263–280. (2000).
19. A. Schoenfeld, Toward a theory of teaching-in-context. *Issues in Education,* **4**, 1, pp. 1–94. (1998).
20. A. Schoenfeld, J. Minstrell, Jim, & E. van Zee. The detailed analysis of an established teacher carrying out a non-traditional lesson. *Journal of Mathematical Behavior,* **18**, 3, 281–325. (2000).
21. A. Schoenfeld, A highly interactive discourse structure, in J. Brophy, Ed., *Social Constructivist Teaching: Its Affordances and Constraints* (Vo!ume 9 of the series *Advances in Research on Teaching*) (New York, Elsevier, 2002), pp. 131–170.
22. A. Schoenfeld. On Modeling Teachers' In-The-Moment Decision-Making, in A. Schoenfeld, Ed., *A study of teaching: Multiple lenses, multiple views.* (A volume in the *Journal for research in Mathematics Education* monograph series). (Reston, VA, National Council of Teachers of Mathematics, in press.).
23. A. Arcavi & A. Schoenfeld. Using the Unfamiliar to Problematize the Familiar. In *TendÍncias Internacionaios em FormaÁao de Profesores de Matem·tica (International Trends in Teacher Preparation),* M. Borba, Ed. (Sao Paulo, Brazil, AutÍntica, 2006), pp. 87–111.

CHAPTER 4

BELIEFS: WHAT LIES BEHIND THE MIRROR?

Gilah C. Leder
La Trobe University—Australia

> *What a man (sic) believes upon grossly insufficient evidence is an index into his desires—desires of which he himself is often unconscious. If a man is offered a fact which goes against his instincts, he will scrutinize it closely, and unless the evidence is overwhelming, he will refuse to believe it. If, on the other hand, he is offered something which affords a reason for acting in accordance to his instincts, he will accept it even on the slightest evidence. The origin of myths is explained in this way.*
>
> —Bertrand Russell

INTRODUCTION

Some years ago, Günter Törner and I, together with Erkki Pehkonen, edited a book—*Beliefs: A hidden variable in mathematics education?* (Leder, Törner, & Pehkonen, 2002). Of the reviews it received, I was particularly taken with the one by John Mason (2004) who set the context for the book as follows:

> The book arose from a working conference held in Oberwolfach in 1999. The task was to come to grips with beliefs and their role in the teaching and learning of mathematics. The first stumbling block is to work out what beliefs

The Montana Mathematics Enthusiast, pages 39–54

actually are, and where they fit into an entire alphabet of associated interlinked terms:

> *A is for attitudes, affect, aptitude, and aims; B is for beliefs; C is for constructs, conceptions, and concerns; D is for demeanor and dispositions; E is for emotions, empathies, and expectations; F is for feelings; G is for goals and gatherings; H is for habits and habitus; I is for intentions, interests, and intuitions; J is for justifications and judgements; K is for knowing; L is for leanings; M is for meaning-to; N is for norms; O is for orientations and objectives; P is for propensities, perspectives, and predispositions; Q is for quirks and quiddity; R is for recognitions and resonances; S is for sympathies and sensations; T is for tendencies and truths; U is for understandings and undertakings; V is for values and views; W is for wishes, warrants, words, and weltanschauung; X is for xenophilia (perhaps); Y is for yearnings and yens; and Z is for zeitgeist and zeal.* (Mason, 2004, p. 347)

Though presumably written at least in part "tongue in cheek," Mason's creative beliefs-alphabet provided a strong impetus for a more detailed look at current research on beliefs—a topic of undoubted interest to Günter. Two recently held annual conferences, those of the *International Group for the Psychology of Mathematics Education* [*PME*][1] and of the *Mathematics Education Research Group of Australasia* [*MERGA*][2] served as convenient and timely data sources of recent research.

Conference proceedings, Leder and Grootenboer (2005) confirmed, capture a variety of research endeavors. They typically include findings from masters and doctoral work, from pilot studies, as well as data from larger projects. These are shared with the mathematics education research community in a variety of formats. For the 2007 PME conference refereed activities comprised Research Forums, Research Reports, Short Oral presentations, and Poster Presentations. For a limited number of sessions more active attendees' participation was encouraged in Discussion Groups and Working Sessions. Invitational plenary activities—Lectures and a Plenary Panel—were also part of the program. A similar, though slightly narrower, range of activities was offered at the 2007 MERGA conference.

A refereeing process is used for both conferences. For example, for the PME conference:

> The Programme Committee received 180 RR [Research report] papers for consideration. Each full paper was blind-reviewed by three peer reviewers, and then these reviews were considered by the Programme Committee, a committee composed of members of the PME international mathematics education community.... In general if there were three or two recommendations for accept, the paper was accepted.... Of the 180 proposals we received, 109 were accepted ... [as research reports]. (PME31 Programme Committee, 2007, p. 1–xliii)

A similar approach was used for MERGA conferences:

> All research papers and symposia submitted were blind peer-reviewed (without the author/s being identified), by two experienced mathematics education researchers who followed strict guideline that have been honed over a number of years. Where the two reviewers, who did not know the identity of the other reviewer, disagreed about the acceptability of a paper, another blind review was carried out by a third reviewer. For consistency, a small panel of highly experienced reviewers undertook the task of reviewing papers in this category. Only those research papers that were accepted by two reviewers have been included in these conference proceedings. (Beswick & Watson, 2007, p. iii)

For both conferences Short Oral Communications were also refereed but their acceptance was based on less stringent rules.

Thus, collectively, the Research Reports[3] and Short Oral Communications and their written representations offered a comprehensive overview of recent work judged by peers to be of a worthy standard in the mathematics education research community. The extent to which research on beliefs forms part of that body of work is discussed in the remainder of the paper. Indicative guidelines for reflecting on the quality of the research draw particularly on Hopkins and Antes' (1990) observation that educational research "focuses on the solutions of problems" and "adds to what is known by providing new knowledge" (p. 23) as well as on Romberg's (1992) list of research activities and in particular "build(ing) a tentative model" and "relat(ing) the phenomenon and model to others' ideas" (p. 51).

THE STUDY

The study comprised the following tasks:

1. Identify work presented as Research Reports and Short Oral Communications at the MERGA and PME conferences in 2007 in which issues related to beliefs and mathematics education were mentioned.
2. Describe the methods used to identify or measure beliefs.
3. For papers in which beliefs were considered in some depth, summarize the belief-related findings.

Method

The proceedings for both conferences are available in both hard copy and CD format. The latter readily enables a search to be made for articles which contain the words believe(s) or belief(s). As already indicated, the search focused on Research Reports and Short Oral Presentations.[4] The latter were included because of the large number of such presentations at the 2007 PME conference.

RESULTS

The PME Conference Proceedings

One hundred and five Short Oral Presentations and 109 Research Reports were included in the PME conference proceedings.

SHORT ORAL PRESENTATIONS

Despite the restricted word length imposed on contributions in this category many authors conveyed the scope of their research effectively.

Close to 20% of the Short Oral Presentations contained a reference to believe or beliefs. However, in the majority of these the term seemed to be used as a stylistic device or to sketch some background information through a brief reference to previous literature rather than as part of a serious attempt to engage with "beliefs." Examples of such stylistic, non technical usage included:

- "I believe that the study makes a strong case for how increasing and enhancing classroom interaction through task-based teaching can help to foster students' cognitive skills and overall mathematical ability" (Leung, 2007, 1–258),[5] and
- "In conclusion, the researcher believed that, in an effort to enhance the professional development, (the) teacher needs to explore an innovative teaching technology" (Phachana & Kongtaln, 2007, 1–274).

Context setting examples included

- "Research has found that the beliefs teachers have regarding mathematics, including their level of mathematical confidence, have a significant impact on their practice of teaching, and hence on the confidence of their students" (Burgoyne, 2007, 1–201); and
- "Learning to teach can be seen as a process of identity formation that brings together one's past experiences, present beliefs and future possibilities" (Prescott & Cavanagh, 2007, 1–276).

TABLE 4.1 Short Oral Presentations with a Strong Focus on Beliefs

Author	Study focus	Measurement	Findings included
Krzywacki-Vaino & Pehkonen	Exploration of teacher identity by focusing on one pre-service teacher	Semi structured interviews and some written material (Subject's work)	Feelings and beliefs about teaching mathematics and being a teacher differ.
Rughubar-Reddy	Students' beliefs and attitudes towards mathematical literacy and values transmitted by the teacher. One male and one female student were the key informants	Interviews, questionnaires, journals, and classroom observations	The two students differed in the values they acknowledged. Their beliefs were reflected in their classroom behaviors.
Sumpter	Influence of beliefs on mathematical reasoning used in problem solving	Video records of students' solving of tasks, interviews, and questionnaires	Safety/security beliefs, expectations and motivations, rather than mathematical knowledge, influenced strategy choice.

In only three of the Short Oral communications did beliefs appear to be the central focus of the research presented. These, summarised in Table 4.1, relied on a small sample and used multiple instruments to tap participants' beliefs. The strong reliance on qualitative, high inference measures is noteworthy.

RESEARCH REPORTS

Approximately half of the Research Reports, 55, referred to beliefs. Despite the more generous word length permitted for Research Reports, in the majority reference to beliefs was again restricted to stylistic usage or to setting the context for the research described in the rest of the paper. Two examples are given below:

- A paper headed *Elementary education students' memories of mathematics in family context* (Hannula, Kaasil, Pehkonen, & Laine, 2007) contained a brief reference to parental and children's beliefs in the introductory section. In the general description of the project, students' "beliefs" was among the variables hypothesized to be predictive of mathematics achievement. Yet there was no further mention of "beliefs" *per se* in the remainder of the paper. Instead participants' narratives dwelled on getting help, role models, value of mathematics, encouragement/demands, independence, and helping siblings."

- In their research report *What is a beautiful problem? An undergraduate students' perspective* Koichu, Katz, and Berman (2007, 3-113) described "investigat(ing) high school students, undergraduate students and mathematics teachers' beliefs and actions through the lens of mathematical aesthetics..." as one of the aims of their project. Yet beliefs were rarely mentioned in the rest of the report. The "theoretical background section" contained a brief reference to beliefs. In a whole group discussion, students were asked to express "their beliefs about what a beautiful problem is." The rest of the paper was silent on beliefs. Instead the authors used words like opinion, perception, perspective, conception, and self esteem.

In both these papers, as well in many others in which the links between beliefs and behaviors are acknowledged, the authors' conceptualization of beliefs remained elusive and intangible.

Those papers, 14 (i.e., 13%), in which belief-related concerns comprised a core component, and in which attempts are made to measure beliefs, are summarized in Table 4.2. Table entries focus on the "belief" aspects even if other issues were also covered in the paper.

The MERGA Conference Proceedings

Eighty-one Research Reports and 28 Short Communications were presented at the MERGA conference.

SHORT ORAL PRESENTATIONS

Only two of the Short Communications mentioned beliefs. Because of the severe word length constraints imposed on the written format of these presentations it is difficult to grasp the scope of the research presented from the contents of the proceedings alone. Thus no further analyses were attempted.

RESEARCH REPORTS

Approximately half (46) of the Research reports referred to belief or believe; of these 11, i.e., 14% of all the research papers, had a sustained focus on beliefs. These latter reports are summarized in Table 4.3. As for the PME

TABLE 4.2 Research Reports with a Strong Focus on Beliefs, PME 2007 Conference

Author	Study focus	Measurement	Findings included
Baturo, Cooper, & Doyle	Professional development (PD) program for 11 indigenous mathematics teacher assistants.	Observations during the PD program, interviews, and a 5-point rating scale which included assessing "affects and beliefs"	Participants' affects, "motivation and confidence" about mathematics and mathematics tutoring improved.
Chapman	Use of inquiry approaches by secondary pre-service mathematics teachers, N = 2	Interviews, classroom observations, and teaching documents	Beliefs about mathematics and beliefs about students' learning contributed to the participants' "sense making of using inquiry approaches."
Ng, Stillman, & Stacey	Effect of interdisciplinary project work on Singapore students' perceptions of mathematics, N = 409 students aged 12–14.	5-point Likert scales administered before and after specially devised project work	Perceptions of interconnectedness comprised two broad factors: inter-subject learning and beliefs and efforts in making connections. Some gender and stream differences were found.
Forgasz & Mittelberg	Comparison of 215 grade 9 Israeli Jewish and Israeli Arab students' beliefs about the gender stereotyping of mathematics	Mathematics as a Gendered Domain instrument	The Arab Israeli students' gendered views of mathematics were ambiguous and differed from those of the Jewish Israeli students and grade 9 students in Australia and the USA.
Halverscheid & Rolka	Eliciting mathematical beliefs through pictures and associated text	Ratings of drawings created by 5th grade students "to express their views about mathematics"	With training, high inter-rater reliability was achieved coding the pictures as reflecting "instrumentalist," "Platonist," and "problem-solving" views of mathematics.
Kapetanas, & Zachariades	Beliefs and attitudes about studying mathematics, N = 1645 10–12th graders	28 item questionnaire, 10 of which probed beliefs (and 14 focused on attitudes)	Two of the five factors identified concerned beliefs: about "utility of proofs and mathematics" and "mathematical understanding through procedures." Positive correlations were found between these factors and students' performance.

(continued)

TABLE 4.2 Research Reports with a Strong Focus on Beliefs, PME 2007 Conference (continued)

Author	Study focus	Measurement	Findings included
Markovits & Pang	Comparison of Korean and Israeli students' beliefs about mathematics (as well as their performance on mathematical tasks), N = 275 grade 6 students	Six questionnaire items about beliefs. These items were not included in the paper	Korean students preferred exact calculations; Israeli students were more likely to rely on number sense. In the body of the paper reference to beliefs was implicit rather than explicit.
Melo & Pinto	Exploration of affective processes through which one student's "self-image as a mathematics learner" is developed, N = 1	Attitude questionnaire, written task—a movie script about mathematics, semi structured interviews, and classroom observations	Intensive interrogation of the data enabled the researchers "to identify beliefs which are grounded in her socio-cultural context and in her relations with others, as well as feelings built during her life experience with mathematics … in the classroom". An experienced teacher was not able to modify the student's negative mathematics-related self-image as a learner.
Novotná & Hošpesová	Possible link between the Topaze effect ("spoon feeding" students towards the correct answer) and teachers' beliefs (and influence of this on students' work) N = one 8th grade class	Videotaping of lessons and post-lesson video-stimulated interviews	Evidence that teachers believed their prompts were needed for students to complete assigned tasks successfully.
Olson, Olson, & Okazaki	Parents' competence beliefs for their children's success in mathematics and their children's "self-efficacy and interest in mathematics", and language used when solving mathematical tasks, subset of N = 66 students and 44 parents	Parallel parent and child survey focusing on self-efficacy, value/usefulness, and competency beliefs; videotapes	Major focus on development and validation of instruments and coding procedures. Gender differences were found: "the mother asked more perceptual questions that … require one-word responses whereas the father asked more conceptual questions that focused on relationships and more abstract ideas."

Perger	Link between espoused theory (what students say is best practice) and theory in use (what they do), N = Pasifika 11–13 year olds in one large co-educational school	Semi-structured interviews exploring beliefs about successful mathematics learning; classroom observations	Students believed that they used practices valued and promoted by the teacher: listening to the teacher, listening to others, having time to think, working with others, and asking for teachers' clues. High and low achieving students differed in their understanding of what these practices entailed.
Rolka, Rösken, & Liljedahl	Change in beliefs as conceptual change, N = 39 pre-service primary school teachers	Reflective journal in which participants documented beliefs. Three prompt questions were given.	Whether changes in beliefs occurred depended on the presence of cognitive conflict (identical/approximate/incomplete fit) between the participants' existing beliefs and their course experiences.
Sakonidis & Klothou	Assessment of pupils' written work in mathematics, N = 553 primary school teachers	Assessment of four 10 year old students' solutions to word problems requiring operations with whole numbers.	Teachers' beliefs about the nature of mathematics included: "mathematics is the route to the result," "the student's thought is mathematically logical," "the student uses the shortest route and this shows intelligence and correct mathematical thinking."
Wang & Chin	Intervention by mentors in the mathematics teaching of pre-service teachers, N = 8 mentor-practice teachers & their students	Classroom observations focusing on critical incidents of teaching, pre- and post-lesson interviews, mentor–tutor conferences	A mentor's knowledge, beliefs and experiences about (teaching and) mentoring influence intervention decisions and follow-up mentoring strategies. Mentors experience their own critical incidents.

TABLE 4.3 Research Reports with a Strong Focus on Beliefs, MERGA 2007 Conference

Author	Study focus	Measurement	Findings include
Bailey	Narrative inquiry by one pre-service teacher educator to investigate her professional practice	Researcher's journal entries, pre-service teachers journals, audiotapes, interviews and observations	"Valuable personal learning" took place: previously unrecognized beliefs about learning and the nature of mathematics were identified; beliefs and teaching practice changed.
Brady	Link between teachers' role, learners, Learning and mathematics, and prospective teachers' views of their ideal primary mathematics classroom, N = 22	Analysis of 750–1000 word report by students of their personal philosophy of teaching primary mathematics, including a description of an ideal primary mathematics classroom.	Idealized views of learners and learning dominated the descriptions of the imaginary classroom. Classrooms were going to be "safe havens," mathematics was "fun and enjoyable," and lessons would be varied, "interactive and relevant."
Cavanagh & Prescott	Exploration of factors which influence the pre-service teachers' classroom practice, N = 8	Three clinical interviews about practicum experiences, including major influences, lesson style, and (in)consistencies with pedagogical focus of the course	Participants identified a mismatch between reform teaching approaches advocated in the university course and observed classroom practices. The researchers gained "the distinct impression that the participants were telling us what they thought we wanted to hear rather than what they really believed".
Frid & Sparrow	Factors influencing the implementation of innovative nontraditional mathematics teaching and learning practices, N = 8	Interview	The recent graduates (1–4 years since graduation revealed "how aspects of their pre-service education provided them with the knowledge, skills, and confidence to enact their beliefs about effective mathematics teaching."
Goos & Bennison	Interactions between teachers' knowledge and beliefs, their professional contexts, and their formal and informal PD experiences, N = 2	Semi structured, theory-guided interview, 40 item questionnaire, and videotaped classroom observations	Valsiner's (1997) zone theory, which builds on Vygotsky's Zone of Proximal Development by adding descriptors of the social setting and the goals and actions of participants, was useful for studying the "interactions between teachers' knowledge and beliefs, their professional contexts, and their formal and informal PD experiences."

Goos, Dole, & Makar	Examination of teacher learning through evaluating the effectiveness of a theoretically based PD model, N = 4 pairs of teachers	Whole group interview, 40 item Mathematics Belief questionnaire, lesson notes, student work, and classroom observations	Different configurations of *teacher knowledge and beliefs* (Zone of Proximal Development), *professional contexts* (Zone of Free Movement), and *sources of assistance* (Zone of Promoted Action) were identified. How these factors combined to shape opportunities for teacher learning was also illustrated.
Ingram	In depth study of mathematical identities and negative affective responses to mathematics, N = one 14-year-old student.	Audiotaped and videotaped observations, interviews, student and parent autobiographical questionnaires, an anxiety questionnaire, student work, journal entries, achievements, and subject choices.	Capturing the student's multiple mathematical identities provided a context for understanding his affective responses in mathematics. His "rare, but increasing, instances of negative affective responses can be seen as a result of a gap between his designated and actual identities."
McNaught	Mathematics autobiography of a student who showed no mathematics anxiety, N = 1	Mathematics autobiography	An autobiography can provide insights for teachers and teacher educators into students' classroom experiences and beliefs about mathematics teaching and learning. The study also highlighted the value of reflective writing as a tool for pre-service teachers' "self-growth."
Rogers	An examination of the relationship between professional learning, classroom practices, and teacher beliefs and attitudes, N = 1 teacher	25 item survey, observations, video footage, and written comments and reflections by the participating teacher	"Guskey's (1986) argument that it is when teachers use new ideas *and* gain evidence of positive change that a change may occur in their beliefs" is supported. His model is more effectively considered to be cyclic rather than linear.
Skalicky	Link between teachers' beliefs and practices and student learning, N = 2 grade 8 teachers and 6 students	Teacher and student semi-structured interviews, and samples of student work	The teachers' beliefs about numeracy and its value and role in the curriculum were able to be explored through the interviews. These beliefs were further reflected in the discourse of the six students in this study.
Wilson	Exploration of bibliotherapy as a reflective tool in teacher education by analyzing affective responses of pre-service primary teachers studying an elective number theory unit, N = 11 pre-service primary teachers	Description of participants' critical learning incident and guided reflections on mathematics education readings which collectively addressed psychological and sociocultural aspects of learning mathematics and focussed on affective and cognitive factors.	"The juxtaposition of bibliotherapy with mathematics teacher education units has proved to be a powerful strategy to address mathematics anxiety in pre-service teachers." The "catharsis, insight, universalisation, and projection allow(ed) the pre-service teachers to reflect more coherently on their beliefs about mathematics learning and teaching."

proceedings, papers in which aspects of beliefs featured in the literature review but were not a focal point of the research described are omitted from the table.[6]

SUMMARY

Papers Containing References to Beliefs

The similarity in the numbers of research papers in the two sets of proceedings in which belief-related issues were addressed is striking: in both cases approximately half referred to beliefs. Of these, 13% and 14% of the PME and MERGA research papers respectively focused in more detail on beliefs and indicated how beliefs were tapped or measured. In the remainder the term "beliefs" was typically used generically or as a synonym for one of the words in Mason's list reproduced earlier in the paper. The papers in the latter group did not add substantively to the body of research on beliefs. Nor, however, did they reflect Bloom's (1981) warning issued almost three decades ago that "in education, we continue to be seduced by the equivalent of snake-oil remedies, fake cancer cures, perpetual-motion contraptions, and old wives' tales. Myth and reality are not clearly differentiated, and we frequently prefer the former to the latter (p. 15).

Measurement/Instruments Used

Approaches to the measurement of beliefs were diverse: Likert questionnaires, semi-structured interviews, observations with or without audio/videotape backing, and reflective writings were strongly favoured. Reliance on interview or reflective writing excerpts to capture participants' beliefs was common. Studies in which the author(s) clearly elucidated the theoretical underpinnings of the analyses adopted were relatively rare. Studies in which neurological or physical phenomena were used to probe participants' beliefs, increasingly prevalent in the wider research community, were glaringly absent.

Yet there were a number of interesting attempts to go beyond popular measures. These included ratings of student drawings and assessment of their reliability, production of a movie script about mathematics, use of a mathematics biography, and an adaptation of bibliotherapy as ways of tapping beliefs about various aspects of mathematics.

Issues Explored

How beliefs influenced teacher practices and student learning, were formed or changed—often as a result of participation in a professional development program or pre-service course activities—attracted considerable research attention. Small sample/in-depth or case study/multiple instrument studies predominated. The latter observation mirrors the findings of Adler et al. (2005) who concluded after their extensive review of research on mathematics teacher education that there was "a notable absence of large scale studies" (p. 370). For the work presented at the PME and MERGA conferences, it should be added that reference was made in a number of reports to a larger research study of which the material presented at the conference was a only a subset.

CONCLUDING COMMENTS

Inspection of the 2007 PME and MERGA conference proceedings confirmed the mathematics education research community's continuing interest in the ways students' and teachers' beliefs affect mathematics learning and instruction, the persistent usage of belief / believe as a convenient synonym for a host of other words, and the frequent failure to distinguish carefully and consistently between beliefs and other affective factors. Small sample, qualitative methods studies, of questionable generalizability, were particularly favored. The extent to which such studies generate new theories, test the limits of a specific theory, "add to what is known by providing new knowledge" (Hopkins & Antes, 1990, p. 23) or constructively relate "the phenomenon and model to others' ideas" (Romberg, 1992, p. 51) is far from clear.

At the same time, examination of the two sets of conferences proceedings identified studies with painstakingly crafted, theoretically driven explorations and creative projective techniques. These not only exemplified Hopkins and Antes' (1990) description of educational research as "a structured scientific inquiry into an educational question that provides an answer contributing toward increasing the body of generalizable knowledge about educational concerns" (p. 24) but point to promising new avenues for capturing more effectively students' and teachers' multi-faceted beliefs, and their impact on mathematical learning and behaviors.

NOTES

1. The 31st annual PME conference was held in Seoul, Korea, July 8–13, 2007.
2. The 30th annual MERGA conference was held in Hobart, Tasmania, Australia, July 2–6, 2007.
3. PME uses the terminology Research Reports and Short Oral Communications. For MERGA conferences the equivalent terminology is Research Papers and Short Communications. Corresponding terms are used interchangeably in this paper.
4. Initially I had intended to focus only on Research Reports. However, because of the exceptionally large number of Short Oral Presentations at the 2007 PME conference I was concerned that much relevant information might be missed if these presentations were ignored.
5. Reference details are given at the end of the paper. They conveniently capture the many different topics within which belief-related concerns were embedded.
6. As already mentioned, a strong emphasis on beliefs in developing the context of the study did not necessarily imply that there was "sustained" focus on beliefs in the research aspect described in the paper.
7. To avoid unnecessary repetitions, editors' names and full title of the proceedings have been omitted from individual entries.

REFERENCES

Adler, J., Ball, D., Krainer, K., Lin, F-L, & Novotna, J. (2005). Reflections on an emerging field: Researching mathematics teacher education. *Educational Studies in Mathematics, 60*, 359–381.

Bloom, BS (1981*). All our children learning.* New York: McGraw-Hill.

Hopkins, C.D. & Antes, R.L. Educational research. Indiana: Peacock Publishers.

Leder, G. C., & Grootenboer, P. (2005). Editorial: affect and mathematics education. *Mathematics Education Research Journal, 17*(2), 1–8.

Leder, G.C., Pehkonen, E., & Törner, G. (Eds.). *Beliefs: A hidden variable in mathematics education?* Dordrecht, The Netherlands: Kluwer.

Mason, J. (2004). Are beliefs believable? *Mathematical Thinking and Learning, 6*(3), 343–351.

Romberg, T.A. (1992). Perspectives on scholarship and methods. In D.A. Grouws (Ed.) *Handbook of Research on Mathematics Teaching and Learning* (pp. 49–64). New York: MacMillan.

Watson, J., & Beswick, K. (Eds.) (2007). *Mathematics: Essential research, essential practice. (Proceedings of the 30th annual conference of the Mathematics Education Research Group of Australasia.).* MERGA Inc: Adelaide, South Australia.

Woo, J-H, Lew, H-C, Park, K-S, & Seo, D-Y. (Eds.) (2007). Proceedings of the 31st conference of the International Group for the Psychology of Mathematics Education. Seoul, Korea: The Korea Society of Educational Studies in Mathematics.

Papers from the *2007 PME Conference Proceedings*[7]

Baturo, Annette R. & Cooper, Tom J. & Doyle, Katherine (2007). Authority and Esteem Effects of Enhancing Remote Indigenous Teacher-Assistants' Mathematics-Education Knowledge and Skills, Vol 2. 57–64.

Burgoyne, Nicky (2007). An Enactive Inquiry into Mathematics Education: A Case Study of Nine Preservice Primary School Teachers, Vol 1. 201.

Chapman, Olive (2007). Preservice Secondary Mathematics Teachers' Knowledge and Inquiry Teaching Approaches, Vol 2. 97–104.

Dawn, Ng Kit Ee & Stillman, Gloria & Stacey, Kaye (2007). Interdisciplinary Learning and Perceptions of Interconnectedness of Mathematics, Vol. 2. 185–192.

Forgasz, Helen J. & Mittelberg, David (2007). The Gendering of Mathematics in Israel and Australia, Vol. 2. 233–240.

Halverscheid, Stefan & Rolka, Katrin (2007). Mathematical Beliefs in Pictures and Words Seen through "Multiple Eyes", Vol. 2. 281–289.

Hannula, Markku S. & Kaasila, Raimo & Pehkonen, Erkki & Laine, Anu (2007). Elementary Education Students' Memories of Mathematics in Family Context, Vol. 3.1–8.

Kapetanas, Eleftherios & Zachariades, Theodosios (2007). Students' Beliefs and Attitudes about Studying and Learning Mathematics, Vol. 3. 97–104.

Koichu, Boris & Katz, Efim & Berman, Abraham (2007). What is a Beautiful Problem? An Undergraduate Students' Perspective, Vol. 3. 113–120.

Krzywacki-Vainio, Heidi & Pehkonen, Erkki (2007). Development of Teacher Identity during Mathematics Teacher Studies, Vol. 1. 246.

Leung, King Man (2007).Using classroom interaction to develop students' Mathematical ability and cognitive skills, Vol. 1. 258.

Markovits, Zvia & Pang, JeongSuk (2007). The Ability of Sixth Grade Students in Korea and Israel to Cope with Number Sense Tasks, Vol. 3. 241–248.

Melo, Silvana Martins & Pinto, Marcia Maria Fusaro (2007). Exploring Students' Mathematics-Related Self Image as Learners, Vol. 3. 257–264.

Novotna, Jarmila & Hošpesova, Alena (2007). What is the Price of Topaze? Vol. 4. 25–32.

Olson, Melfried & Olson, Judith & Okazaki, Claire (2007). A Study of Gender Differences in Language Used by Parents and Children Working on Mathematical Tasks, Vol. 4. 49–56.

Perger, Pamela (2007). If You Don't Listen to the Teacher, You Won't Know What to Do: Voices of Pasifika Learners, Vol. 4. 73–80.

Phachana, Preechakorn & Kongtaln, Pasaad (2007). The Process of Mathematics Teacher Changing Role in Classroom Using the Story and Diagram Method, Vol. 1. 274.

Prescott, Anne & Cavanagh, Michael (2007). Pre-service Secondary Mathematics Teachers: A Community of Practice, Vol. 1. 276.

Rolka, Katrin & Rosken, Bettina & Liljedahl, Peter (2007). The Role of Cognitive Conflict in Belief Changes, Vol. 4. 121–128.

Rughubar-Reddy, Sheena (2007). Students' Beliefs and Attitudes in a Mathematical Literacy Classroom, Vol. 1. 278

Sakonidis, H. & Klothou, A. (2007). On Primary Teachers' Assessment of Pupils' Written Work in Mathematics, Vol. 4. 153–160.
Sumpter, Lovisa (2007). Beliefs and Mathematical Reasoning, Vol. 1. 289.
Wang, Chih-Yeuan & Chin, Chien (2007). How Do Mentors Decide: Intervening in Practice Teachers' Teaching of Mathematics or Not, Vol. 4. 241–248.

Papers from the *MERGA 2007 Conference Proceedings*[7]

Judy Bailey (2007). Mathematical Investigations: A Primary Teacher Educator's Narrative Journey of Professional awareness, Vol. 1. 103–112.
Kathy Brady (2007). Imagined Classrooms: Prospective Primary Teachers Visualise their Ideal Mathematics Classroom, Vol. 1. 143–152.
Michael Cavanagh & Anne Prescott (2007). Pre-service Teachers into a Community of Practice, Vol. 1. 182–191.
Sandra Frid & Len Sparrow (2007).Towards "Breaking the Cycle of Tradition" in Primary Mathematics, Vol. 1. 295–304.
Merrilyn Goos & Anne Bennison (2007), Technology-Enriched Teaching of Secondary Mathematics: Factors Influencing Innovative Practice, Vol.1. 315–324
Merrilyn Goos, Shelley Dole, & Katie Makar (2007). Supporting an Investigative Approach to Teaching Secondary School Mathematics: A Professional Development Model, Vol. 1. 325–334.
Naomi Ingram (2007). A Story of a Student Fulfilling a Role in the Mathematics Classroom, Vol. 1. 450–459.
Keith McNaught (2007). The Power of Writing for all Pre-service Mathematics Teachers, Vol. 2. 473–482.
Pauline Rogers (2007). Teacher Professional Learning in Mathematics: An Example of a Change Process. Vol. 2. 631–640.
Jane Skalicky (2007). Exploring Teachers' Numeracy Pedagogies and Subsequent Student Learning across Five Dimensions of Numeracy, Vol. 2. 661–670.
Sue Wilson (2007). My Struggle with Maths May Not Have Been a Lonely One: Bibliotherapy in a Teacher Education Number Theory Unit, Vol. 2. 815–823.

CHAPTER 5

POLITICAL ISSUES IN MATHEMATICS EDUCATION

Ubiratan D'Ambrosio
Pontifícia Universidade Católica de São Paulo, Brazil

For Günter Törner, in honor of his 60th birthday.

INTRODUCTION

Almost ten years ago, I delivered the Paulo Freire Memorial Lecture, at the First Mathematics Education and Society Conference (MEAS1) in Nottingham, UK, September 6–12 1998. I started stating that my views on Mathematics Education reflected my perception that much of the problems modern society faces is reflected in a system of codes of behavior, anchored in current knowledge, which accepts facts and events with a character of "normality". The apparent success and the dominant presence, in both the people and the decision making elites, of the *normopaths*, which are individuals who *a critically* conform to what is normal, very often lead to a generalized hindrance and even repression of creativity, through censoring and silencing individuals. This scenario is seen nowadays.

For decades we have been noticing that the search of new, alternative spaces, by those who have no voice, has been a high motivation for most

The Montana Mathematics Enthusiast, pages 55–62

youth action. As recent examples, I mention the emergence of "off-Broadway" theatre. I could refer to something from the early sixties, the "free-speech movement" led by Lenny Bruce and the students revolts in Paris and in many other cities around the World, in 1968. We can refer also to the proliferation of religious faiths and fundamentalism as the search of alternative for the very "normal" established institutional scenario. This proliferation might be called "off-Church" movement. And the increase in school failings and dropouts result in the "off-School" trend in youth behavior. While in the cases of drama and the churches there are institutionalized alternative spaces, off-School means the streets, teenage pregnancy, drugs, violence and youth criminality.

Much of the search for off-school spaces is a result of the insistence on teaching obsolete, useless and disinteresting subjects. As Simon Jenkins, a millennium commissioner and former editor of *The Times,* puts it, "pupils become very bored [school is disinteresting]...and they do not see science as getting them jobs [science is useless]" *brackets are mine.*[1] Although Jenkins refers to science teaching, the situation with mathematics is considerably worst. We might very well add obsolete. I have many times written disinteresting, obsolete and useless.

It is clear that insisting on the three R's which still prevails in most school systems will not remedy school failings. Another emphasis is required, which focuses on relevant issues of today's society, responds to youth intellectual curiosity and agility, and prepares them for a future, which we can only guess how will it be? Particularly in education, we need a new thinking, which brings peace, justice, respect and solidarity.[2]

In the taped lecture Paulo Freire prepared for ICME 9, he recognized that mathematics is intertwined with all forms of human behavior and that there is a mathematical way of being in life. He essentially recognizes that his program of Critical Literacy cannot be complete without the recognition that mathematics underlies human and societal behavior.[3] This goes much beyond the acquisition of mathematical skills. In the Paulo Freire Memorial Lecture, I discussed, under the inspiration of his ideals, the role of Mathematics in building up a new civilization which rejects inequity, arrogance and bigotry.

POLITICAL ISSUES

Political issues deal with government, economics, relations among nations and social classes, people's welfare, the preservation of natural and cultural resources, and many other sectors of our life. The state of the world causes apprehension. The possibility of final extinction of civilization in Earth is real. Not only through war. We are witnessing an environmental crisis, disruption of the economic system, institutional erosion, mounting social crises in just about every country and, above all, the recurring threat of war.[4]

History shows us that mathematics is well integrated in the technological, industrial, military, economic and political systems and that mathematics has been relying on these systems for the material bases of its continuing progress. It is important to question the role of mathematics and mathematics education in the perverse behavior of mankind.

We have to look into history and epistemology of mathematics with a broader view. Particularly with respect to the cultures of the periphery, with the denial and exclusion of knowledge. This was so common in the colonial process, but still prevails in modern society. The denial of knowledge to individuals, particularly children, is of the same nature as the denial of knowledge which affects entire populations. To propose directions to counteract ingrained practices is the major challenge of educators, particularly mathematics educators. Large sectors of the population do not have access to full citizenship. Some do not have access to the basic needs for survival. This is the situation in most of the world and occurs even in the most developed and richest nations. Our hopes for the future depend on learning—critically—the lessons of the past. A scenario similar to the political disruption of the Roman Empire is before us, with the aggravation that the means of disruption are, nowadays, practically impossible to control.

We may consider survival with dignity the most universal problem. It is also generally agreed that mathematics is the most universal mode of thought. I see the understanding of the relation between these two universals as a step towards actions to avoid the bleak scenario of the future of civilization. Surely, mathematics offers the means to propose new approaches to human and social relations and to the relations with the environment. Consequently, I exhort mathematicians and mathematics educators to reflect about our personal role in looking for the ways our science can justify its advancements as means to achieve survival with dignity for the humanity.

Although specially referring to the threat of Nuclear War, the appeal of the Einstein-Russell Manifesto, of 1955, applies equally well to the current state of the world:

> We appeal, as human beings, to human beings: Remember your humanity and forget the rest. If you can do so, the way lies open for a new paradise; if you cannot, there lies before you the risk of universal death.[5]

The nature of mathematical behavior is not yet clearly understood. Although in Classical Philosophy we can notice a concern with the nature of mathematics, only recently the advances of cognitive sciences have probed into the generation of mathematical knowledge. How is mathematics created? How different is mathematical creativity from other forms of creativity?

HISTORY AND EDUCATION

From the historical viewpoint, there is a need of a complete and structured view of the role of Mathematics in building up our civilization. For this we have to look into the history and geography of human behavior and find new paths in the measure we advance in the search. History is a global view in time and space, and it cannot be focused only on the narrow geographic limits of a few civilizations which have been successful in a short span of time. This limited view favors History as a chronological narrative of events, focused in results and heroes. Consequently, this leads to a biased view of history, which favors interests of groups.

The course of the history of mankind, which can not be separated from the natural history of the planet, reveals an increase interdependence which crosses space and time, of cultures and civilizations and of generations.

Education is a strategy created by societies to promote creativity and citizenship. To promote creativity implies helping people to fulfill their potentials and raise to the highest level of their capability. To promote citizenship implies showing them their rights and responsibilities in society.[6] Educational systems throughout history and in every civilization have been focusing on two issues: to transmit values from the past and to promote the future. In other words, education aims equally at the new (creativity) and the old (societal values).

But we should be very concerned with irresponsible creativity—for we do not want our students to become bright scientists creating new instruments of oppression and mass destruction—and with docile submission to societal values—for we do not want our students to accept rules, norms and codes which violate human dignity. This is our challenge as educators, particularly as mathematics educators.

The strategy of educational systems to pursue these goals is the *curriculum*. Curriculum is usually organized linearly, in three strands: objectives, contents and methods. Thus, the curriculum emerges from the acceptance of the social aims of educational systems, then the identification of contents which may help to reach the goals and, finally, in the development of methods to transmit these contents.

THE POLITICAL DIMENSION OF MATHEMATICS EDUCATION

The political dimension of mathematics education must be analyzed in the three strands of the curriculum: objectives, contents, methods. The political discourse about mathematics education normally focuses on the objectives. Very rarely has mathematics contents and methodology been examined under this dimension. Some educators and mathematicians claim that contents and methods in mathematics have nothing to do with the political dimension of education.

Even more disturbing is the possibility, in a world convulsed by wars, of conveying to the children war is unavoidable, so they must be uncritically prepared for it. Since Mathematics is the imprint of the Western thought, it is naïve not to look into a possible role of mathematics in framing a state of mind which tolerates war. Our responsibility as mathematicians and mathematics educators is to offer venues of peace.[7]

There is an expectation about our role, as mathematicians and math educators, in the pursuit of peace. The remark below by Anthony Judge, the Director of Communications and Research of the Union of International Associations, express how we, mathematicians, are seen:

> Mathematicians—having lent the full support of their discipline to the weapons industry supplying the missile delivery systems—would claim that their subtlest thinking is way beyond the comprehension of those seated around a negotiating table.[8]

I see my role as an Educator and my discipline, Mathematics, as complementary instruments to fulfill commitments to mankind. In order to make good use of these instruments, I must master them, but I also need to have a critical view of their potentialities and of the risk involved in misusing them. This is my professional commitment.

It is easily recognized that Mathematics provides an important instrument for social analyses. Western civilization entirely relies on data control and management. Social critics will find it difficult to argue without an understanding of basic mathematics. But, regrettably, the term "basic" has been abusively identified with *a critical* skill and drilling.

It is an undeniable right of every human being to share all the cultural and natural goods needed to her/his material survival and intellectual enhancement. This is the essence of the *Universal Declaration of Human Rights* (1948), to which every nation is committed. The educational strand of this important profession on mankind is the *World Declaration on Education for All* (1990), to which 155 countries are committed. Of course, there are many difficulties in implementing the effectiveness of the United Nations resolutions and mechanisms. But as yet this is the best instrument available that may lead to a planetary civilization, with Peace and dignity for the entire mankind. Regrettably, Mathematics Educators are generally unfamiliar with these documents.

It is an un-relinquishable duty to cooperate, with respect and solidarity, with all the human beings, who have the same rights, for the preservation of all these goods. This is the essence of the *ethics of diversity*: respect for the other (the different); solidarity with the other; cooperation with the other. This leads to quality of life and dignity for the entire mankind. It is impossible to accept the process of exclusion of large sectors of the population of the World, both in the developed and undeveloped nations. An explanation for this perverse concept of civilization asks for a deep reflection on the colonial period. It is not the case of putting the blame in one or another, neither to attempt to redo the past. But to understand the past is a first step to move into the future. To accept inequity, arrogance and bigotry is irrational and may lead to disaster. Mathematics has everything to do with this state of the world. A new world order is urgently needed. Our hopes for the future depend on learning—critically!—the lessons of the past.

We have to look into history and epistemology with a broader view. The denial and exclusion of the cultures of the periphery, so common in the colonial process, still prevails in modern society. The denial of knowledge which affects populations is of the same nature as the denial of knowledge to individuals, particularly children. To propose directions to counteract ingrained practices is the major challenge of educators, particularly of mathematics educators. Large sectors of the population do not have access to full citizenship. Some do not have access to the basic needs for survival. This is the situation in most of the world and occurs even in the most developed and richest nations.

In order to build up a civilization which rejects inequity, arrogance and bigotry, education must give special attention to the redemption of peoples that have been for a long time subordinated and must give priority to the empowerment of the excluded sectors of societies.The Program Ethnomathematics contributes for restoring cultural dignity and offers the intellectual tools for the exercise of citizenship. It enhances creativity, reinforces cultural self-respect and offers a broad view of mankind. In everyday life, it is a system of knowledge which offer the possibility of a more favorable and harmonious relation in human behavior and between humans and nature.[9] A consequence of this program for a new curriculum is synthesized in my proposal of three strands in curricular organization: Literacy, Matheracy and Technoracy.[10]

> **Literacy:** Clearly, reading has a new meaning today. We have to read a movie or a TV program. It is common to go to a concert announced as a new reading of Chopin! Also socially, the concept of literacy goes through many changes. Nowadays, "reading" includes also the competency of numeracy, interpretation of graphs, tables and other ways of informing the individual. And also understanding the condensed language of codes. These competencies have much more to do with screens and button than with pencil and paper. There is no way for reverting this trend, the same as there was no successful

censorship in preventing people to have access to books in the last 500 years. Getting information through the new media precedes the use of pencil and paper and numeracy is dealt with calculators. But, if dealing with numbers is part of modern literacy, where has mathematics gone?

Matheracy: is the capability of drawing conclusions from data, inferring, proposing hypotheses and drawing conclusion. It is a first step towards an intellectual posture, which is almost completely absent in our school systems. Regrettably, even conceding that problem solving, modeling and projects can be seen in some mathematics classrooms, the main importance is given to numeracy, or the manipulation of numbers and operations. Matheracy is closer to the way Mathematics was present both in classical Greece and in Indigenous cultures. The concern was not with counting and measuring, but with divination and philosophy. Matheracy, this deeper reflection about man and society, should not be restricted to the elite, as it has been in the past.

Technoracy: is the critical familiarity with technology. Of course, the operative aspects of it are, in most of the cases, inaccessible to the lay individual. But the basic ideas behind the technological devices, their possibilities and dangers, the morality supporting the use of technology, are essential questions to be raised among children in a very early age. History shows us that ethics and values are intimately related to technological progress.

The three together constitute what is essential for conscious citizenship in a world moving fast into a planetary civilization.

NOTES

1. *The Times*, London, 8/09/98, p. 8
2. See the interesting book by Richard A. Slaughter (ed.): *New Thinking for a New Millenium*, Routledge, London, 1996.
3. A conversation with Paulo Freire, *For the Learning of Mathematics,* vol. 17, n. 3, November 1997, pp. 7–10.
4. See Sriraman, B. (2007). On the origins of social justice: Darwin, Freire, Marx & Vivekanda. In B. Sriraman (Ed), *International Perspectives on Social Justice in Mathematics Education*, Monograph 1, The Montana Mathematics Enthusiast, pp. 1–6.
5. http://www.pugwash.org/about/manifesto.htm accessed July 30, 2007.
6. D'Ambosio, U. (2007). Peace, Social Justice, Ethnomathematics. In B. Sriraman (Ed), *International Perspectives on Social Justice in Mathematics Education*, Monograph 1, The Montana Mathematics Enthusiast, pp. 25–34.

7. Ubiratan D'Ambrosio: Mathematics and peace: Our resposibilities, *Zentralblatt für Didaktik der Mathematik/ZDM,* Jahrgang 30, Juni 1998, Heft 3; pp. 67–73.
8. Anthony Judge: And when the bombing stops? Territorial conflicts as a challenge to mathematicians, http://www.uia.org/uiadocs/mathbom.htm accessed July 30, 2007.
9. Ubiratan D'Ambrosio: Ethnomathematics and its First International Congress, *Zentralblatt für Didaktik der Mathematik,* ZDM 99/2; pp. 50–53.
10. Ubiratan D'Ambrosio: Literacy, Matheracy, and Technoracy: A Trivium for Today, *Mathematical Thinking and Learning,* 1(2),1999; pp. 131–153.

CHAPTER 6

SNAPSHOTS FROM THE 1960s

Tensions and Synergies in the Emerging of New Trends in Mathematics Education

Fulvia Furinghetti
Dipartimento di Matematica dell'Università di Genova, Italy

To Günter Törner

Je n'aspire pas à changer la société, mais malgré des désillusions que j'ai éprouvées, je crois toujours qu'on peut changer peu à peu l'enseignement [...].

—H. Freudenthal, as quoted in (Adda, 1993, p. 12)[1]

INTRODUCTION

In September 1997 a school teacher in California posed to the web Discussion List on the History of Mathematics[2] the following question: "What happened in mathematics in the year 1968?" The reason for this question was an interdisciplinary theme weekend on the year 1968 planned in his school (grades 9–12, academically strong). I have recorded four answers. One men-

The Montana Mathematics Enthusiast, pages 63–81

tioned the final computations made and programmed to send a manned vehicle from the earth to its moon, and then return. The second pointed out that 1968 was the year of the famous month-long Global Analysis summer school at Berkeley. The writer also evoked vividly the particular climate of the hippie period in this University. For another respondent 1968 was the year in which V. Kac and R. Moody independently introduced and classified the class of Lie algebras now called Kac-Moody algebras. The last responder resorted to St Andrew Archive[3] and found that in 1968 the following known mathematicians died: S. N. Bernstein, J. F. A. Delsarte, A. O. Gelfond, L. D. Landau (1962 Nobel Prize for physics), K. Loewner, L. Roth.

It is my opinion that an analogous question referring to mathematics education ("What happened in mathematics education in the year 1968?") would have as a first answer "the foundation of the journal *Educational Studies in Mathematics.*" In this paper I outline some elements that are behind this event and make this journal a milestone in the history of mathematics education. The recent developments, the policy, and the place in the international panorama of *Educational Studies in Mathematics* have been outlined and discussed when the fiftieth volume was issued and when the centenary of its predecessor journal *L'Enseignement Mathématique* was celebrated, see (Furinghetti et al., 2003; Hanna, 2003; Hanna & Sidoli, 2002). The focus of the present paper is in the factors, which raised the need of founding such a journal. My main sources will be journal itself and *L'Enseignement Mathématique*; moreover I'll use some data gathered in my study on activities relating to mathematics education in the proceedings of the quadrennial International Congresses of Mathematicians (see Furinghetti, to appear). The question underlying the present paper is the tormented relation between mathematicians and mathematics educators. Which better subject to celebrate Günter Törner, who was and is so successful in dealing with this relation?

L'ENSEIGNEMENT MATHÉMATIQUE, ICMI, IMU, ICM

The journal *Educational Studies in Mathematics* does not come out of the blue: its roots and antecedents rely in the history of the journal *L'Enseignement Mathématique* and in the links of this late journal with ICMI.

L'Enseignement Mathématique was founded in 1899 by the French C. A. Laisant and the Swiss H. Fehr. As discussed in Furinghetti (2003), the aim of this journal was to foster communication, internationalization and solidarity in mathematical instruction around the world. The modern settlement of various nations was completed almost everywhere and there was the need to build instructional systems. The information on national systems of instructions and the comparison of curricula were important in such a construction and the journal promoted this policy of information: in this way it sowed the seed for the creation of ICMI (International Commission on Mathematical Instruction), which happened in the International Congress of Mathematicians of Rome (6-11 April 1908). The ideas of communication and internationalization were spreading also in the world of the professional mathematicians. Laisant himself, director with É. Lemoine of the journal *L'Intermédiaire des Mathématiciens* launched in this journal (1894, *1*, question 212, p. 113) the idea of an international congress of mathematicians. As we know this idea was well received by the mathematicians and the first International Congress of Mathematicians took place in Zurich (9-11 August 1897). This short historical outline[4] evidences a particular link of the world of mathematical instruction/education and the world of mathematicians. ICMI (until the 1950s mainly indicated with the French acronym CIEM or the German IMUK[5]) was born inside the International Congress of Mathematicians and the mandates for its future activities and its committees were decided in connection with these Congresses. Thus the life of ICMI paralleled the life of the community of mathematicians. Wars had been shocking events in these lives. Wars, indeed, are storms that swap away existing situations. This, in particular, may happen to international institutions, which, for their very nature, are based on values such as solidarity, spirit of collaboration and wish of communicating that wars put in crisis. The life of ICMI was signed by the two World Wars. For example, after the Versailles Peace Treaty the dissolutions of international scientific associations provoked the first lethargy of ICMI, which lasted until the resurrection in the ICM-1928 (Bologna). The Second World War stopped again the activities and the new resurrection happened in 1952 with a new name and acronym—International Mathematical Instruction Commission (IMIC)—that soon was changed into the present name International Commission on Mathematical Instruction (ICMI). In the ICM-1954 (Amsterdam) ICMI was lively present, see (Furinghetti, to appear). Since then the new life of ICMI developed towards the present pattern of a pathway with lights and shadows.

Two strong elements of continuity characterize the period 1908-1954: *L'Enseignement Mathématique* and H. Fehr. Since the foundation of ICMI the journal was its official organ: it published reports of the various countries, announcements and information, minutes of ICMI meetings, ICMI inquiries. H. Fehr was Secretary General of ICMI until 1952; in 1954 (the year of his death) he was appointed as honorary president of ICMI. It is unanimously acknowledged that he has been the real soul of ICMI, the indefatigable ferryman of the international mathematical instruction world from the nineteenth century to the twentieth century.

H. Fehr's death marked the end of an era: society was changing and a new generation of ICMI members was emerging. Mathematics itself was undergoing changes in paradigms. At the ICM-1954 in Amsterdam (the second after the war) J. A. Schouten in his presidential address drew the attention to "a fact which was perhaps not so clear four years ago, but which is absolutely clear now: *the place of mathematics in the world has changed entirely after the second war*" (Gerretsen & de Groot, 1957, Vol. 1, p. 143); moreover he stressed the increasing importance of "our modern computing machines" (ibidem, p. 145). The new role of mathematics was promoted by advances in technology and the pressures by governments to give an education adequate to the changed setting of the world. The concern for the iron curtain was pushing towards strong developments in technology and mathematics instruction was perceived as an important factor of power for the nations.

In the first General Assembly held in Rome (9 March 1952) the International Mathematical Union had reconstituted the old CIEM/IMUK under the ephemeral name of "International Mathematical Instruction Commission (IMIC)". In the general assembly of IMU (The Hague, 1 September 1954) the present name International Commission on Mathematical Instruction and the acronyms ICMI were adopted again (even if for some period translated into French in the reports published in *L'Enseignement Mathématique*). As witnessed by these reports ICMI continued its activity of inquiring and organized meetings often sponsored jointly by other bodies. The proceedings of many of these meetings were published.

After the Second World War several international organizations came into being. With at least three of them ICMI had ties by virtue of common memberships—namely, with the Commission Internationale pour l'Étude et l'Amélioration de l'Enseignement des Mathématiques, the Inter-American Committee on Mathematical Education, the Committee on Mathematics in South Asia.

There were some forms of cooperation with UNESCO—namely organization of conferences in co-operation or consultation of members, publications of reports and books. Several members of ICMI made important individual contributions to Royaumont Seminar in 1959 sponsored by OEEC[6] and on the work on its Dubrovnik Report in 1960. A formal request was made by OEEC to ICMI in 1959 for preparing a list of outstanding mathematical texts. For a long time there has been a semi-formal understanding between ICMI and the Musée Pédagogique for the mounting of the exhibitions of books and other educational materials at the International Congresses of Mathematicians.

In spite of all these activities in the sixties ICMI was losing its centrality as regards mathematics education. In the report (International..., 1963) we see some signs of difficulties in the relationship with mathematicians. An evidence of these difficulties is the SCOTS (Special Committee on the Teaching of Sciences) affair. The affair is briefly described in the 'Report for the period 1959–1962' (International..., 1963, p. 110) as follows:

> In 1960 U.N.E.S.C.O.'s Department of Natural Sciences begun discussing with I.M.U. the possibilities of formal co-operation in the field of mathematical instruction at the university level. These discussions resulted in the conclusion of a contract for this purpose between U.N.E.S.C.O. and I.M.U. early in 1962, but the executive committee of I.M.U. decided to create a Special Committee on the Teaching of Science (S.C.O.T.S.) to handle its obligations under the contract as well as its developing general interests in the broader field of science education. A close co-operation between I.C.M.I. and S.C.O.T.S. is thus to be desired in the future.

It is likely that ICMI would have wished that SCOTS had been appointed as a sub-commission of ICMI. This episode is a step in the long-lasting history of frictions between IMU and ICMI. The point was, in the very words of Lehto (1998, p. 110)

> The Executive Committee of the IMU had mixed feelings about the steps the Commission had taken. On one hand, the activity of the Commission was welcomed. But the Executive Committee wished to exercise some control over its sub-commission, which was supposed to be a link between research mathematicians and teachers and which did not possess financial resources of its own.

The frictions with the community of mathematicians are surprising if one considers that the mathematicians, F. Klein for one, were the main supporters in the first period of ICMI. But times had changed since then. The ICMI reports (International..., 1963; Commission..., 1966) show that financial autonomy from IMU was a main concern of ICMI. We wonder that another problem could have been the concern about academic positions: as a matter of fact the fifth of the 'Resolutions of the First International Congress on Mathematical Education', published in the proceedings of ICME-1 (*Educational Studies in Mathematics*. 1969–1970, *2*, 135–418) claims:

> The theory of mathematical education is becoming a science in its own right, with its own problems both of mathematical and pedagogical content. The new science should be given a place in the mathematical departments of Universities or Research Institutes, with appropriate academic qualifications available. (p. 416)

In the ICMI report (1966, pp. 136–137) we read that in the meeting of Paris (15 February 1964) ICMI (with the agreement of IMU) decided to acknowledge the status of national sub-commissions also to some national commissions representative of countries not belonging to IMU. Thus it was officially recognized that ICMI had a far wider target population than did its parent body IMU. In spite of the wide scope of ICMI activities, the proceedings of the ICMs held in the 1960s (1962 in Stockholm and 1966 in

Moscow) dedicated little space to ICMI. *L'Enseignement Mathématique* hosted some reports that were not published in the proceedings of ICMs.

The 1960s were years of great ferment in the world of mathematics education. New curricular projects were springing up in many countries. The debate on modern mathematics was alive. As regards this debate *L'Enseignement Mathématique* published important papers, such as (Freudenthal, 1963; Piaget, 1966), but other arenas existed for this discussion, CIEAEM (Commission Internationale pour l'Étude et l'Amélioration de l'Enseignement des Mathématiques)[7] for one. This commission, officially founded in 1952, had begun its activities in 1950, having C. Gattegno as promoter and animator. CIEAEM was independent from IMU and from national organizations: the members were a group of people (including shool teachers) sharing common objectives as regards mathematics education and working in an atmoshere of friendly relationship. Over the years CIEAEM kept a rather informal character (only in 1996 it had a "constitution") and new members were co-opted. In her short history E. Castelnuovo (1981) says that the aim of the commission was to stress that the mathematicians alone were not sufficient for a deep study of the teaching problem: a wider view was necessary and it required the help of psychologists and pedagogists. She mentions also the influence of the young's movement in 1968 as an element influencing the view of mathematics teaching.

The backbone of CIEAEM activities were the periodical meetings of the commission restricted to the members or open to a larger audience. These latter meetings were organized as international conferences and—with some exceptions—had their proceedings. The two collective publications issued in the pioneering period illustrate the scope of the Commission in its intial years (Gattegno et al., 1958; Piaget et al., 1955). Bernet and Jaquet (1998) report the names of 23 founder members, among them J. Dieudonné (strong supporter of modern mathematics), J. Piaget and two future presidents of ICMI (H. Freudenthal, A. Lichnerowicz). It is remarkable that five of the 19 members of the editorial board of the first issue of *Educational Studies in Mathematics* appear in this list: E. Castelnuovo, L. Félix, G. Choquet, H. Freudenthal, W. Servais.

We may suppose that some of Freudenthal's ideas underlying the foundation of *Educational Studies in Mathematics* developed in the context of CIEAEM, whose meetings he attended: the perplexities about modern mathematics; the need of involving in the discusssion of educational problems not only mathematicians, but also teachers and psychologists; the need for confronting also ideas, not only programs or curricula.

Freudenthal's views about modern mathematics are discussed in (Adda, 1993; Goffree, 1993). Goffree (1993) feels that at the beginning Freudenthal underestimated the impact of this movement; he also reports that he complained to have not attended the Royaumont Seminar, which prompted the New Math movement in Europe. In (Goffree, 1993, p. 29) we read that, according to Freudenthal "Too many mathematicians, who had nothing whatsoever to do with mathematics education, have been invited."

The cooperation with teachers was advocated by Freudenthal already in the programs for the period 1955–1958, as reported in (Commission..., 1955, p. 200):

> Freudenthal pense qu'il faudrait que ces enquêtes suscitent un travail en profondeur; il serait souhaitable, par example, que des professeurs de l'enseignement secondaire puissent faire connaître leurs essais, les résultats de leurs expériences; une étude comparée pourrait alors s'instituer avec des données précises.[8]

About the object of research in mathematics education Freudenthal (1963, p. 29) wrote:

> L'histoire a démontré la stérilité des problèmes d'organisation pure. Dans ces dernières années l'accent c'est porté sur les programmes. C'est inquiétant, cette activité des programmeurs. À maintes reprises j'ai insisté sur les recherches franchement didactiques. Il est vrai que jusqu'alors le résultat de mes efforts est assez maigre.[9]

NEW NEEDS IN MATHEMATICS EDUCATION

As told before the death of H. Fehr (2 November 1954) closed an era. It is announced in the last volume of the first series (1951-1954, *40*, p. 3). A brief note (ibidem, p. 4) to subscribers and readers signed J. Karamata, J. Piaget, G. de Rahm informs that the last fascicule of *L'Enseignement Mathématique* had been prepared by Fehr and was published on demand of Fehr's widow. Starting from 1955 the journal would be quarterly. This last issue of the first series kept the early days format with some adaptations (see Furinghetti, 2003), namely the following sections:

- general articles (methodology and various notes, history);
- the organization of instruction;
- report on Swiss Mathematical Society;
- chronicle;
- bibliographic bulletin;
- the list of authors.

There is also a list of the papers and the reports of ICMI published in the first 40 volumes.

The reports of ICMI appear in French and English (Commission..., 1951–1954, pp. 72–93). We read (p. 81): "At its first General Assembly, held in Rome from 9 March 1952, the International Mathematics Union reconstituted The International Mathematical Instruction Commission (IMIC) by calling upon H. Behnke (Munster, Westpalia), A. Chatelet (Paris), H. Fehr (Geneva), R. L. Jeffery (Canada), D. Kurepa (Zagreb, Yugoslavia)". The Commission met in Geneva (21–22 October 1952), in Paris (21 February 1953 and 15 January 1954). Additional members were co-opted to form the executive committee. At page 82 of this volume we read: "In his circular dated 7 May 1953, Prof. Châtelet of C.I.E.M. proposed to take: *L'Enseignement Mathématique*, well known by all readers, as the Commission's official publication."

The first meeting of the new ICMI Executive Committee took place in Geneva (2 September 1955). In the same year the first volume of the second series was issued. In the note 'The review "L'Enseignement Mathématique"' (Commission..., 1955, pp. 270-271) there is the presentation"

> This second series will be devoted to the reform and development of mathematical instruction; it will publish articles focusing on and explaining modern theories in a manner comprehensible to non-specialized mathematicians; deal with arrangement and organization of teaching; study the psychological formation of mathematical ideas; and publish accounts of the work done and surveys carries out by the I.C.M.I. A bibliographical index will be included in each number.

The financial issues were not clearly settled by the International Mathematical Union, however "the I.C.M.I. deems it its duty to support this publication by doing its best to increase the numbers of subscribers" (*s. 2, 1*, p. 271). With this editorial statement the new life of *L'Enseignement Mathématique* began. The journal continued to be the official organ of ICMI, but slowly its role became less significant outside the Francophone countries. There was a tension between the use of the two languages—French and English—in the journal: ICMI report for the period 1959–1962 (M. H. Stone President) published in 1963 (*s. 2, 9*, pp. 105–112) is in English, that of the period 1963–1966 (A. Lichnerowicz President) published in 1966 (*s. 2, 12*, pp. 131–137) is in French with the old acronym of CIEM.

The problem of language is only one reason of the journal's decline as an educational organ. As a matter of fact, on the one hand, gradually its format changed into the usual format of mathematical journals and often the published papers had no relation with instruction/education. On the other hand, the developments of society and school made the mere study and comparison of curricula and programs—which was the initial objective of the journal—inadequate to face the complexity of the educational problems. In the title of the short lecture delivered by L. N. H. Bunt (from the

Institute of Education of Utrecht) at the ICM-1954 in Amsterdam there is the new expression "didactical research", which reveals the emerging orientation in mathematics education. The decline of ICMI and the problematic relation with mathematicians were other reasons of decline.

The minutes of the ICMI session (in the journal still CIEM) held in Utrecth (26 of August 1967) gave the situation of ICMI just before the creation of *Educational Studies in Mathematics* (Commission..., 1967). The points approached in the meeting are summarized below:

- Information on the conference of Lausanne sponsored by UNESCO in January 1967 (see below the first volume of *Educational Studies in Mathematics*).
- A. Revuz announced the publication of the first volume of *New Trends*, the series published by UNESCO with the collaboration of ICMI. The following editorial board was proposed: A. Revuz (president), A. Z. Krygowska, H. G. Steiner, D. H. Wheeler, J. Suranyi, M. Glaymann, H. G. Fehr.
- Announcements of:
    - A 12-day conference in October 1968 in Bucharest (Rumania) sponsored by Rumania and UNESCO. About 30 participants were expected.
    - A conference on the integration of the various sciences at secondary level continuing the conference of Lausanne organized by CIES (the commission for integration of sciences in teaching). CIES was discussing with UNESCO the publication of a work on the applications of mathematics to sciences.
- Various proposals for the future activities of ICMI coming from the delegates of various nations present at the meeting.
- The President expressed disappointment about the modality of ICMI participation at the quadrennial Congresses of Mathematicians, which was confined to the presentation of reports. Generally, the national reports were unusable. The President supported the idea of a congress of ICMI to be held a year before the ICM, where invited talks and personal communications could be presented. The assembly accepted the project of a Congress of ICMI in 1969. The French delegate M. Glaymann proposed to hold the congress in France.
- The creation of a center of documentation of ICMI. According to Steiner such a center was already active in Southern Illinois University and he himself was co-director for Europe.
- A. Revuz asked for a new journal closer to secondary teachers, because *L'Enseignement Mathématique* has at too high a level. There were different proposals and at the end the assembly decided the creation of a commission for studying this question. "Behnke,

Châtelet, Hilton, Novak, Pescarini, Steiner, Thwaites ainsi que M. and Freudenthal" were appointed as members (p. 245).

- It is proposed to restrict the treatment of school organization and the programs to the meeting and conferences of ICMI. For the next meeting of Bucharest the themes proposed were the following: mathematization, motivation, how to teach mathematics without the teacher, comparative evaluation of the contents of mathematical courses, issues fostering the success, evaluation of the results of research of mathematics education, methodology of research.
- Contact with developing countries
- Proposals for reviving the national commissions (increasing the number of members at large, creation of a permanent secretariat).

THE FOUNDATION OF A NEW JOURNAL, THE NEW TRADITION OF ICME CONFERENCES

As president of ICMI (1967–1970) Freudenthal took two important initiatives aimed at raising the profile of ICMI. Firstly, since he did not see *L'Enseignement Mathématique* as meeting the needs of the day, in 1968 founded *Educational Studies in Mathematics,*[10] as a publication independent from ICMI. Obviously, this action provoked a friction with IMU. Lehto (1998, p. 259) reports that

> At the meeting of the IMU Executive Committee held in Paris in May 1968, President [H.] Cartan and Secretary [O.] Frostman complained of the lack of information about the activities of ICMI. [...] The Executive Committee had not been told of the creation by ICMI of the new journal *Educational Studies in Mathematics*, which seemed to compete with *L'Enseignement Mathématique.* A financial contract had been signed between ICMI and UNESCO without the IMU having been informed.

The second initiative taken by Freudenthal was the creation of ICME Congresses, a permanent institution which after the second ICME in 1972 was arranged regularly every four years between, see Appendix 2. With this initiative Freudenthal recovered the tradition, established in the first years of ICMI before the First World War, of having international congresses detached from the ICMs. The first ICME was held in Lyon (France), during 24–30 August 1969, thanks to financial subventions from the French government and UNESCO. The initiative was received by IMU with coldness, as evidenced by the sentence "it seems that ICMI decided to hold an international congress in Paris in 1969" reported in (Lehto, 1998, p. 259) from the *ICMI-Bulletin of the International Commission on Mathematical Instruction* (January 1984, *n. 15*, pp. 17–20).

Both the initiatives (*Educational Studies in Mathematics* and ICME) were shocking events in the life of ICMI and of *L'Enseignement Mathématique.* There is no mention of ICMI and of the new journal in the volume 14 (1968, *s. 2, 14*).[11] The entire issue of 1969 (no number series indicated, *15*) is dedicated to J. Karamata (1902–1967), without notes on ICMI or on the Lyon conference. We may note that, in spite of these events that could have been appeared as a tear with the past, in his address at ICME-1 Freudenthal (1969–1970) acknowledged the scope without borders of ICMI's activities since the beginning of the century.

The first volume (1968–1969) of *Educational Studies in Mathematics*[12] does not contain an editorial statement. However in his address Freudenthal (1968–1969, p. 3) provides a list of propositions on the teaching of mathematics referring to the resolutions[13] adopted in the meeting of Lausanne (January 1967) sponsored by UNESCO, which was attend also by physicists. These resolutions are published in this first volume (pp. 244–246), together with recommendations about the coordination of the teaching of mathematics and physics. The eight points of the resolutions concern:

- the right of all children to be educated through mathematics, a unique and characteristic activity of the human mind.
- the development of the capacity of intellectual action instead of merely piling up of knowledge
- the knowledge and the mastership of mathematical structures and its utilization in the grasp of reality as objectives of mathematics teaching
- the attempt to use these structure from childhood on
- the acquisition of some more sophisticated structures until the end of secondary school
- the training of teachers
- the re-training of teachers in relation with the permanent process of reformation of mathematics teaching. The retraining has to be based on regular pedagogical research
- the global collaboration in this field and the "urgent requirement to establish an international organism for information on the teaching of mathematics" (p. 244).[14]

A synthesis of some important Freudenthal's ideas comes from the title of the Colloquium held in Utrecht, 26 of August 1967: "How to teach mathematics so as to be useful." This colloquium was "an activity of ICMI sponsored by the government of the Netherlands and by IMU." The Proceedings are published in the first volume of *Educational Studies in Mathematics* (pp. 1–243).

The authors in the first volume were members of the editorial board or eminent chief characters in the world of education of that period. The languages used were English, French, and German. We remark the concern for the relation of mathematics with other disciplines and the presence of two papers (by M. Glaymann and by R. J. Walker) on the use of computers.

In the counter-cover there is the announcement of the first ICME to be held in Lyon. The second volume contains the addresses and the resolutions (in English and French) of the Congress (1969–1970, pp. 135–418). Thus the proceedings of ICMI-1 were published in two forms: as a book and as a part of the volume 2 of the journal. The Editorial Board of *Educational Studies in Mathematics* is the editor of the proceedings. In the allocution to the Congress Freudenthal (1969–1970) alternates English and French language.

The Editorial Board (with the only exception of W. T. Martin who was absent from volume 7 onwards) remained unchanged until volume 9 (1978). The new editor A. Bishop constituted "a new Board, independent from the old one"[15] in charge from volume 10 of 1979. Since the foundation of the journal to this year many events important for mathematics educators community had taken place. In 1977 the first conference of IGPME (International Group for the Psychology of Mathematical Education) took place in Utrecht (29 August–2 September). Freudenthal, who had joined the Group in Karlsruhe (ICME-3) accepted to deliver the welcome address, though claiming that "I could hardly define in which way I am interested in psychology" (1978a, p. 1). The volume 9 of *Educational Studies in Mathematics* hosts the texts of talks delivered in this conference. At pages 141–142 of this volume we find the announcement of the approval by ICMI Executive Committee of the affiliation of the "International Study Group on Relations Between the History and Pedagogy of Mathematics, cooperating with the International commission on Mathematical Instruction" (p. 141).

In volume 9 of *Educational Studies in Mathematics* Freudenthal published a special issue (two numbers, pp. 147–379) containing the national reports on the theme "Changes in mathematical education since the late 1950s—ideas and realisation. An ICMI report." In the rich variety of approaches, Freudenthal (1978b, p. 145) identified the following "common lesson learned by all concerned in the process of innovations [...]: better understanding of the part played by the teacher in the course of change."

In the "Tribute" published in volume 9 (1978, p. 504) the retiring Editorial Board acknowledged that "*ESM* have neither started nor blossomed had it not been for his [Freudenthal's] initiative, imagination, enthusiasm and energy."

EPILOGUE

The intertwining of the scope of IMU, ICMI, *Educational Studies in Mathematics*, *L'Enseignement Mathématique* had as a central question the relationship between mathematicians and mathematics educators. The tear provoked by the foundation of *ESM* had the merit of fostering a clarification of the domains of actions of these two communities.

L'Enseignement Mathématique pursued a different direction in respect to *Educational Studies in Mathematics*: it became a journal publishing works on pure mathematics, while continuing to be the official organ of ICMI. In 1983 (*s. 2, 29*) the ICMI President J. P. Kahane and the ICMI secretary G. Howson joined the editorial board of *L'Enseignement Mathématique*, which then started publishing discussion documents for the newly-instituted ICMI Studies in its pages.[16] Then old links were renewed and strengthened.

In the allocution at a first ICME Freudenthal (1969–1970, p. 136) claimed that "Mathematics should not be taught to fit a minority, but to everybody, and they should learn, not only mathematics but also what to do with mathematics." Which better way to revive and update the old ideals expressed in the first issue of *L'Enseignement Mathématique* in the light of the new world view promoted by the flower children's revolution in the 1960s?

NOTES

1. I do not aspire to change society, however, in spite of some disillusions I felt, I always trust that little by little we may change teaching. [In the present paper the translations are mine].
2. math-history-list@enterprise.maa.org
3. http://www-groups.dcs.st-and.ac.uk./~history/index.html
4. The history of IMU and its links with the related commissions is in (Lehto, 1998). For the history of ICMI see the website built on the occasion of the celebration for the centenary of ICMI < http://www.dm.unito.it/icmi/>.
5. CIEM stands for Commission Internationale de l'Enseignement Mathématique, IMUK stands for Internationale Mathematische Unterrichtskommission.
6. OEEC (Organisation for European Economic Co-operation) later became OECD (Organisation for Economic Co-operation and Development).
7. International Commission for the Study and Improvement of Mathematics Teaching. This commission is still active (there was only an interuption from 1967 to 1969). It is usually indicated with the French acronym. For the history of CIEAEM see (Bernet & Jaquet, 1998).
8. Freudenthal thinks that it should be needed that the inquiries raise a work in depth; for example, it should be desirable that secondary teachers may let know their works, the results of their experiments; then a comparative study could be organized based on precise data.

9. History has shown the sterility of the problems of mere organization. In the recent years the attention has been directed to the programs. This activity of programmers is worrying. Repeatedly I insisted on studies actually didactical. It is true that until now the result of my efforts is very poor.
10. The timeline of the foundation of *ESM* is in Appendix 1.
11. In the volumes of 1968 and 1970 the year of the volume is different from the year of the publication.
12. For the publisher and the editorial board of the volume 1 see Appendix 3.
13. "In my opinion the resolutions adopted in Lausanne are a milestone in the philosophy of mathematical education. If I substitute my wishes and hopes for my opinion, I would say they should be so" (Freudenthal, 1968–1969, p. 3).
14. These words are reported literally from the text of the resolutions and show some mistrust in the effectiveness of ICMI's action.
15. (Freudenthal, 1978c, p. 504).
16. The first discussion document ('The influence of computers and informatics on mathematics and its teaching') was prepared by R. F. Churchhouse, B. Cornu, A. P. Ershov, A. G. Howson, J. P. Kahane, J. H. van Lint, F. Pluvinage, A. Ralston, M. Yamaguti. It is published in *L'Enseignement Mathématique* (1984), *s. 2, 30,* 161-172.
17. From (Freudenthal, 1978c).

REFERENCES

Adda, J. (1993). Une lumière s'est éteinte—H. Freudenthal homo universalis. *Educational Studies in Mathematics, 25,* 9–19.

Bernet, T. & Jaquet, F. (1998). *La CIEAEM au travers de ses 50 premières rencontres.* Neuchâtel: CIEAEM.

Castelnuovo, E. (1981). CIEAEM: histoire de cette Commission. In M. Pellerey (Ed.). *Proceedings of the 33rd CIEAEM's meeting* (*Processes of geometrisation and visualization,* Pallanza, Italy), 11–14.

Commission Internationale de l'Enseignement Mathématique (1951–1954). *L'Enseignement Mathématique, 40,* 72–93.

Commission Internationale de l'Enseignement Mathématique (C.I.E.M.) (1955). *L'Enseignement Mathématique, s. 2, 1,* 195–202.

Commission Internationale de l'Enseignement Mathématique (C.I.E.M. ou I.C.M.I.) (1966). Rapport sur la période 1963–1966. *L'Enseignement Mathématique, s. 2, 12,* 131–138.

Commission Internationale de l'Enseignement Mathématique (1967). Compte rendu de la séance de la C.I.E.M. tenue à Utrecht, le 16 août 1967. *L'Enseignement Mathématique, s. 2, 9,* 243–246.

Editors and advisory editors of ESM (2002). Reflections on Educational Studies in Mathematics. *Educational Studies in Mathematics, 50,* 251–257.

Freudenthal H. (1963). Enseignement des mathématiques modernes ou enseignement modernes des mathématiques?. *L'Enseignement Mathématique, s. 2, 9,* 28–44.

Freudenthal H. (1968-1969). Why to teach mathematics so as to be useful. *Educational Studies in Mathematics, 1,* 3–8.

Freudenthal H. (1969-1970). Allocution du premier congrès international de l'enseignement mathématique Lyon 24–31 août 1969. *Educational Studies in Mathematics, 2*, 135–138.

Freudenthal H. (1978a). Address to the first conference of the I.G.P.M.E. (International Group for the Psychology of Mathematical Education). *Educational Studies in Mathematics, 9*, 1–5.

Freudenthal H. (1978b). Changes in mathematical education since the late 1950s—ideas and realisation. An ICMI Report. *Educational Studies in Mathematics, 9*, 143–145.

Freudenthal H. (1978c). Acknowledgment. *Educational Studies in Mathematics, 9*, 503–504.

Furinghetti, F. (2003). Mathematical instruction in an international perspective: the contribution of the journal *L'Enseignement Mathématique*. In D. Coray, F. Furinghetti, H. Gispert, B. R. Hodgson & G. Schubring (Eds.). *One hundred years of L'Enseignement Mathématique*, Monographie n. 39 de *L'Enseignement Mathématique*, 19–46.

Furinghetti, F. (to appear). Mathematics education and ICMI in the proceedings of the International Congresses of Mathematicians. *Revista Brasileira de História da Matemática Especial nº 1—Festschrift Ubiratan D'Ambrosio*—(December 2007).

Gattegno, C., Sevais, W., Castelnuovo, E. Nicolet, J. L., Fletcher, T. J., Motard, L., Campedelli, L. Biguenet, A., Peskette, J. W. & Puig Adam, P. (1958). *Le matériel pour l'enseignement des mathématiques*. Neuchâtel: Delachaux & Niestlé.

Gerretsen, J. C. H. & de Groot, J. (Eds.) (1957). *Proceedings of the International Congress of Mathematicians*. Groningen: E. P. Noordhoff N. V.; Amsterdam: North-Holland. Vol. 1.

Goffree, F. (1993). HF: Working on mathematics education. *Educational Studies in Mathematics, 25*, 21–49.

Hanna, G. (2003). Journals of mathematics education, 1900–2000. In D. Coray, F. Furinghetti, H. Gispert, B. R. Hodgson & G. Schubring (Eds.). *One hundred years of L'Enseignement Mathématique*, Monographie n. 39 de *L'Enseignement Mathématique*, 67–84.

Hanna, G. &. Sidoli, N. (2002). The story of ESM. *Educational Studies in Mathematics, 50*, 123–156.

International Commission on Mathematical Instruction (1963). Report for the period 1959–1962. *L'Enseignement Mathématique, s. 2, 9*, 105–115.

Lehto, O. (1998). *Mathematics without borders: A History of the International Mathematical Union*. New York: Springer Verlag.

Piaget, J. (1966). L'initiation aux mathématiques, les mathématiques modernes et la psycholgie de l'enfant. *L'Enseignement Mathématique, s. 2, 12*, 289–292.

Piaget, J., Beth, E. W., Dieudonné, J., Lichnerowicz, A., Choquet, G. & Gattegno, C. (1955, 2° ed. 1960). *L'enseignement des mathématiques*. Neuchâtel: Delachaux & Niestlé.

APPENDIX 1
TIMELINE OF THE FOUNDATION OF *EDUCATIONAL STUDIES IN MATHEMATICS*[17]

- Final session of the ICMI Colloquium on Modern Curricula in Secondary Mathematical Education, held 19–22 December 1964 at the Mathematical Institute of Utrecht State University. The following resolution was adopted (p. 503):

  > The participants in the Utrecht Colloquium on Modern Curricula in Secondary Education feel the urgent need for more international information on national activities in mathematical education, which could be organized and spread by an active and accessible international center of information or by a high level periodical on mathematical education.

  The proceedings were not published. It was necessary to create suitable conditions so that this fact will not happen also at the Utrecht Colloquium in 1967.
- The beginning passage of the circular of 13 April 1965 to people active in mathematical education (p. 503):

  > In order to carry out a resolution adopted at a Colloquium on Modern Curricula in Secondary Education, held at Utrecht in December 1964, I have taken steps to arrive at publishing a high level international periodical on mathematical education. The publishers Reidel at Dordrecht (Netherlands) appear to be favourably inclined towards such a project. (p. 502)

- 21–25 August 1967: Utrecht Colloquium. Freudenthal sought and found Reidel's helpful cooperation before the actual constitution of the Editorial Board of *ESM.* The proceeding will be published in the first volume of *ESM.* Freudenthal (p. 503) writes "In this phase I experienced valuable help from Peter Hilton who advised me about the constitution of the Board of Editors and who gave the journal its present name *Educational Studies in Mathematics.*
- 2 December 1967: Letter of invitation to people to become members of the Editorial Board.
- May 1968: date of the first issue of *ESM (numbers 1/2).*

APPENDIX 2
PROCEEDINGS OF THE CONFERENCES ICME

ICME-1, Lyon (France), 24–30 August 1969

The Editorial Board of Educational Studies in Mathematics (Eds.) (1969). *Actes du premier Congrès International de l'Enseignement Mathématique (Commission Internationale de l'Enseignement Mathématique, CIEM)/ Proceedings of the first International Congress on Mathematical Education (International Commission on Mathematical Education (sic), ICMI).* Dordrecht: D. Reidel. One volume: 286 pages.

Also published in *Educational Studies in Mathematics.* (1969–1970), *2(2/3),* 135–418.

ICME-2, Exeter (UK), 29 August–2 September 1972

Howson, A. G. (Ed.) (1973). *Developments in Mathematical Education. Proceedings of the Second International Congress on Mathematical Education.* Cambridge: Cambridge University Press. One volume: IX + 318 pages.

ICME-3, Karlsruhe (Germany), 16–21 August 1976

Athen, H. & Kunle, H. (Eds.) (1977). *Proceedings of the Third International Congress on Mathematical Education.* Karlsruhe: *Zentralblatt für Didaktik der Mathematik.* One volume: 400 pages

The outcome of the thirteen "sections" of the Congress was collected in a book published by UNESCO (1979). *New trends in mathematics teaching,* Vol. IV. Prepared by the International Commission on Mathematical Instruction.

ICME-4, Berkeley (USA), 10–16 August 1980

Zweng, M., Green, T., Kilpatrick, J., Pollak, H. & Suydam, M. (Eds.) (1983). *Proceedings of the Fourth International Congress on Mathematical Education.* Boston: Birkhäuser. One volume: 725 pages.

ICME-5, Adelaide (Australia), 24–30 August 1984

Carss, M. (Ed.) (1986). *Proceedings of the Fifth International Congress on Mathematical Education.* Boston—Basel—Stuttgart: Birkhäuser. One volume: 401 pages.

ICME-6, Budapest (Hungary), 27 July–3 August 1988

Hirst, A. & Hirst, K. (Eds.) (1988). *Proceedings of the Sixth International Congress on Mathematical Education.* Budapest: János Bolyai Mathematical Society. One volume: 397 pages.

ICME-7, Quebec (Canada), 17–23 August 1992

Gaulin, C., Hodgson, B. R., Wheeler, D. H. & Egsgard, J. (Eds.) (1994). *Proceedings of the 7th International Congress on Mathematical Education.* Québec: Les Presses de l'Université Laval. One volume: 495 pages.

Robitaille, D., Wheeler, D. H. & Kieran, C. (Eds.) (1994). *Selected Lectures from the 7th International Congress on Mathematical Education.* Québec: Les Presses de l'Université Laval. One volume: 370 pages.

ICME-8, Seville (Spain), 14–21 July 1996

Alsina, C., Alvarez, J. M., Niss, M., Pérez, A., Rico, L. & Sfard, A. (Eds.) (1998). *Proceedings of the 8th International Congress on Mathematical Education.* Seville: SAEM Thales. One volume: 539 pages.

Alsina, C., Alvarez, J. M., Hodgson, B., Laborde, C. Pérez, A. (Eds.) (1998). *8th International Congress on Mathematical Education—Selected Lectures.* Seville: SAEM Thales. One volume: 485 pages.

ICME-9 Tokyo-Makuhari (Japan), 31 July–6 August 2000

Fujita, H, Hashimoto, Y., Hodgson, B. R., Lee, P. Y., Lerman, S. & Sawada, T. (Eds.) (2204). *Proceedings of the Ninth International Congress on Mathematical Education.* Norwell, MA; Dordrecht: Kluwer Academic Publishers, One volume: 430 pages.

ICME-10 Copenhagen (Denmark), 4–11 July 2004

Proceedings to appear.

APPENDIX 3
EDITORIAL BOARD OF EDUCATIONAL *STUDIES IN MATHEMATICS,* 1 (1/2), MAY 1968–1969

Editor: H. Freudenthal, Mathematical institute, University of Utrecht, The Netherlands

Editorial Board:

D. K. Abbiw-Jackson, Kumasi, Ghana
E. G. Begle, Stanford University, CA, USA
Mlle E. Castelnuovo, Rome, Italy
G. Choquet, Paris, France
A. Engel, Stuttgart, [West] Germany
Mme L. Félix, Paris, France
H. B. Griffiths, Southampton, England
P. Hilton, New York, NY, USA
C. Hope, Worcester, England
Mme A. Z. Krygovska, Cracow, Poland
W. T. Martin, Cambridge, MA, USA
O. Pollak. Murray Hill, NJ, USA
A. Revuz, Paris, France
W. Servais, Morlanwelz, Belgium
S. Sobolev, Novosibirsk, USSR
H. G. Steiner, Karsruhe, Germany
P. Suppes, Stanford University, CA, USA
B. Thwaites, London, England.

Publisher: D. Reidel Publishing company, Dordrecht (Holland)

CHAPTER 7

MATRIZEN ALS DIAGRAMME

Lisa Hefendehl-Hebeker
Universität Duisburg-Essen

All our thinking is performed upon signs of some kind or other, either imagined or actually perceived. The best thinking, especially on mathematical subjects, is done by experimenting in the imagination upon a diagram or other scheme, and it facilitates the thought to have it before one's eyes.

—Peirce, zitiert nach Hoffmann 2006, S. 171.

Viele Beobachtungen bestätigen, dass für zahlreiche Schüler die Möglichkeit des Handhabens konkret hingeschriebener Objekte von außerordentlicher Wichtigkeit ist.

—Törner 1982, S. 323

Abstract: Dieser Artikel betrachtet das Kurskonzept von Artmann und Törner zu Linearen Algebra und Geometrie aus der Perspektive der semiotischen Erkenntnistheorie von Peirce, wie sie von Hoffmann (2005) dargestellt wird. Dies wird exemplarisch am Beispiel des ersten Kapitels ausgeführt. Dabei zeigt sich, dass die gewählte Betrachtungsweise geeignet ist, den Kurs lerntheoretisch genauer zu charakterisieren und das didaktische Potential herauszuarbeiten.

The Montana Mathematics Enthusiast, pages 83–91

EIN KURSKONZEPT ZUR LINEAREN ALGEBRA

Artmann und Törner publizierten 1980 einen Kursvorschlag zur Linearen Algebra in der gymnasialen Oberstufe, in dem das ausführliche Arbeiten in und mit dem Matrix-Vektor-Kalkül im Zentrum steht (Artmann & Törner 1982, 1984; s. auch Törner 1982).

Der Kurs orientiert sich fachlich an eigenen Erfahrungen sowie am Vorbild des Lehrbuches "Linear Algebra and ist Applications" von G. Strang (1976) und folgt methodisch insbesondere dem dort formulierten Motto "explain rather than deduce," in dessen Sinne die Autoren bestrebt sind,

- Fragen zu stellen, ehe Antworten gegeben werden,
- zuerst Phänomene sichtbar zu machen und danach Erklärungen abzugeben,
- vom Konkreten zum Abstrakten voranzuschreiten und nicht umgekehrt.

Diese Ziele werden sowohl lokal bei der Einführung einzelner Begriffe über Beispiele und Aufgaben umgesetzt wie auch bei der globalen Stoffplanung in der Weise berücksichtigt, dass mit dem "Thema Lineare Gleichungssysteme und Matrizen" ein beziehungsreiches Betätigungsfeld erschlossen und der Begriff des Vektorraumes erst am Schluss des Kurses zur weiteren theoretischen Durchdringung erarbeitet wird.

Dabei werden die folgenden inhaltlichen Akzente gesetzt:

1. Der Einstieg erfolgt über lineare Gleichungssysteme und Matrizen mit Anwendungen vor allem in den Wirtschaftswissenschaften. Damit eignet sich das Aufgabenmaterial auch für die Vielzahl von Schülerinnen und Schülern, die wenig Verständnis für physikalische Anwendungen mitbringen. In diesem Kontext wird der Matrix-Vektor-Kalkül geometrie-unabhängig eingeführt und es wird ausführlich in und mit ihm gearbeitet. Dabei werden numerische und algorithmische Komponenten eingeflochten.
2. Mit den erarbeiteten Werkzeugen wird in einem zweiten Schwerpunkt ausführlich räumliche Geometrie betrieben. Vektoren, die als algebraische Objekte eingeführt wurden, können nun im geometrischen Gewand sowohl als Pfeile wie auch als Punkte erscheinen. Damit ergeben sich zwei gleichwertige Sprachebenen, die situationsangepasst gebraucht werden. In der thematischen Entwicklung werden klassische Unterrichtsinhalte der vektoriellen Koordinatengeometrie im $\mathbb{R}^3$ durch die Betrachtung linearer Abbildungen in geometrischer Verankerung ergänzt. Das Thema "Projektionsmatrizen" schließlich ermöglicht eine geometrisch orientierte

Lösungstheorie für lineare Gleichungssysteme mit drei Variablen und spannt auf diese Weise den Bogen zum Einstieg zurück. Damit ist für Grundkurse ein befriedigendes Abschlussniveau gesichert.
3. Das anschließende Kapitel über Grundzüge der Vektorraumtheorie dient der theoretischen Abrundung und Vertiefung für entsprechend leistungsfähige Lerngruppen. Bei der Einführung der abstrakten Begriffe kann auf die zuvor erworbenen reichen Erfahrungen mit Matrizen und Vektoren zurückgegriffen werden.

Für das ausführliche Arbeiten in und mit dem Matrix-Vektor-Kalkül in diesem Kurskonzept führen die Autoren sachliche und didaktische Gründe an.

1. Eine Sachanalyse der endlichdimensionalen Linearen Algebra zeigt, dass sich Matrizen hier als universales Werkzeug verwenden lassen. Fast alle wesentlichen Ergebnisse sind mit ihrer Hilfe zu gewinnen bzw. zu formulieren. Matrizen ermöglichen aufgrund ihrer Universalität umgekehrt die Verzahnung von Gebieten. Sie können insbesondere als verbindendes Element zwischen Arithmetik und Geometrie betrachtet werden und finden auch in der Stochastik maßgebliche Verwendung. Somit können Matrizen in der Oberstufenmathematik einen integrierenden Kern darstellen.
2. Aus didaktischer Sicht stellen Matrizen einen beziehungsreichen Unterrichtsgegenstand dar, der zugleich über eine gewisse Handfestigkeit verfügt. "Gerade im Grundkurs hat man hier die Möglichkeit, Schülern ein zwar nicht triviales, aber doch beherrschbares Routinematerial anzubieten, welches zudem noch in sich reichhaltig und interessant ist. Die Erfahrung zeigte, dass die Möglichkeit des Handhabens konkret hingeschriebener Objekte für viele Schüler von außerordentlicher Wichtigkeit ist." (Artmann & Törner 1984, S. 227)

Der Aspekt des "Handhabens konkret hingeschriebener Objekte" verleiht der Thematik in besonderem Maße Züge des diagrammatischen Arbeitens im Sinne der semiotischen Erkenntnistheorie von Ch. S. Peirce, die in der mathematikdidaktischen Diskussion in den letzten Jahren zunehmend Interesse gefunden hat. Eine semiotische Herangehensweise an das Problem des Lernens bietet sich, wie Hoffmann (2001) entfaltet, aus verschiedenen Gründen an. Zum einen ist jede Kommunikation und jede Repräsentation von Wissen auf Zeichen angewiesen. Daher liegt es nahe, die Rolle von Zeichen in Lernprozessen genauer zu untersuchen. Zum anderen könnte die Semiotik einen theoretischen Rahmen bereitstellen, der es erlaubt, konstruktive und rezeptive Momente des Lernens zu verbinden und dadurch zwischen Dichotomien zu vermitteln, die aus mathematikdidaktischer Sicht problematisch sind.

Es erscheint daher reizvoll, das von Artmann und Törner entwickelte Kurkonzept aus semiotischer Perspektive genauer zu betrachten. Aus Zeit- und Platzgründen beschränken wir uns hierbei auf das erste Kapitel. Dazu entfalten wir zunächst Grundzüge einer semiotischen Erkenntnistheorie und orientieren uns dabei an der Darstellung von Hoffmann (2005).

GRUNDZÜGE EINER SEMIOTISCHEN ERKENNTNISTHEORIE

Ch. S. Peirce vertritt eine am Begriff der Tätigkeit orientierte Philosophie der Mathematik. In deren Zentrum stehen der Begriff des Diagramms und das Konzept des diagrammatischen Schließens. Die Rolle von sichtbaren Repräsentationen als Mittel der Erkenntnisgewinnung ist dafür wesentlich.

Diagramme sind die für die mathematische Erkenntnistätigkeit maßgebenden internen oder externen Darstellungen. Dabei kann es sich um geometrische Figuren, Graphen, Formeln oder Matrizen handeln, aber auch ein Satz, ein Urteil oder ein logischer Schluss ist ein Diagramm. Es handelt sich also um einen sehr weit gefassten Begriff, den wir zunächst genauer charakterisieren wollen.

Diagramme werden gemäß den allgemein akzeptierten Regeln eines bestimmten Darstellungssystems konstruiert. Diese können explizit oder unmittelbar ("kollateral" in der Diktion von Peirce) gegeben sein wie z. B. die Alltagssprache, und sie können das Denken präzise definieren oder nur vage bestimmen. Jedenfalls nimmt man zum Zeitpunkt der Diagrammatisierung an, dass diese Regeln ein konsistentes System bilden, d.h. dass sie in einem widerspruchsfreien Zusammenhang stehen. Hierin ist die "Unausweichlichkeit" mathematischen Schließens angelegt. Regeln und Konventionen—z. B. die Syntax einer natürlichen oder künstlichen Sprache oder die Axiome einer mathematischen Theorie—bestimmen auch den Gebrauch von Diagrammen und definieren, welche Transformationen eines Diagramms zugelassen sind und welche nicht. In Form von Inskriptionen, also als "konkret hingeschriebene Objekte" (Törner 1982, s.o.) treten uns Diagramme in materialisierter, wahrnehmbarer Form gegenüber (Dörfler 2006).

Die Entstehungsweise und die Handhabung von Diagrammen haben bestimmte erkenntnistheoretische Implikationen.

Diagramme sind zunächst ikonischer Natur, d.h. sie rufen einen Eindruck von Ähnlichkeit zwischen Zeichen und Bezeichnetem wach. Diese besteht darin, dass in einem Diagramm relationale Strukturen abgebildet werden. Aufgrund ihres konventionellen Charakters sind Diagramme aber mehr als Ikone. Sowohl für den Konstrukteur eines Diagramms wie auch für den Interpreten ist es entscheidend, dass er über die entsprechenden

Konventionen verfügt, die es erlauben, ein Diagramm als solches und nicht nur als ein Ikon zu sehen. In einem Diagramm kann "Wissen von Kulturen" codiert sein.

Insofern Diagramme Relationen repräsentieren, deren Relata variabel sind, also Möglichkeiten repräsentieren, sind sie offen für Interpretationen und erleichtern es, die inhärenten Relationen spielerisch zu verändern – z. B. ein Vorzeichen umzudrehen oder einen zusätzlichen Faktor in Betracht zu ziehen. Diagramme eröffnen somit kreative Spielräume. Sie erlauben Assoziationen und rufen andere ikonisch darstellbare Ideen hervor, die z. B. bei der Lösung eines Problems helfen können.

Nach Peirce ist alles proportional verfasste Denken ein Denken in Diagrammen. Das Experimentieren mit Diagrammen erhält den Charakter logischen Schließens, weil seine Ergebnisse durch die Regeln des Darstellungssystems bestimmt sind. In dieser Erkenntnistätigkeit treffen eine "objektive" und eine subjektive Komponente aufeinander. Die objektive Komponente besteht darin, dass Diagramme eigenständige "semiotische Welten" darstellen, die durch die Rationalität des gewählten Darstellungssystems mehr oder weniger determiniert sind. Die subjektive Komponente besteht darin, dass der Umgang mit den Zeichen und Darstellungssystemen Freiheitsräume und damit kreative Spielräume eröffnet. Diese betreffen die Auswahl und Konstruktion von Darstellungssystemen, Strategien des Experimentierens sowie Möglichkeiten, das System weiter zu entwickeln, neue und unterschiedliche Interpretationen zu finden und es auch metaphorisch zu verwenden, also in neue Kontexte zu übertragen.

Das *diagrammatische Schließen* als die wesentliche mathematische Erkenntnistätigkeit hat dann folgende Schritte:

- ein Diagramm konstruieren;
- mit diesem Diagramm experimentieren, Ergebnisse beobachten und notieren, sich vergewissern, dass ähnliche Experimente ähnliche Resultate haben würden;
- dies in allgemeinen Begriffen zum Ausdruck bringen.

Die *Möglichkeiten der Erkenntnisentwicklung* ergeben sich aus der Interpretation und Transformation vorhandener und der Entwicklung neuer Darstellungsmittel:

- Die Entfaltung von Implikationen gegebener Darstellungsmittel macht sichtbar, was verborgen in den Diagrammen schon angelegt ist und erzeugt damit abgeleitetes Wissen.
- Die kreative Veränderung des Darstellungssystems selbst, z. B. durch Hinzufügen neuer Mittel oder durch Schaffen einer neuen systematischen Ordnung, erzeugt weitergehendes Wissen.

Dabei verwendet jeder Forschungsprozess drei *Grundformen des Schließens*:

- Die *Abduktion* generiert eine neue Idee bzw. findet eine Hypothese angesichts einer erklärungsbedürftigen Tatsache. Sie geschieht spontan, "ohne Zwang," und enthält ein wesentlich kreatives Moment im Erkenntnisprozess.
- Die *Deduktion* ist notwendiges Schließen. Sie leitet aus der abduktiv gewonnenen Theorie Konsequenzen ab. Auch diese Schlussform enthält kreative Momente, sofern sie Ideen einsetzt, um verborgenes Wissen hervorzuholen.
- Die *Induktion* sucht die experimentelle Bestätigung der abgeleiteten Implikationen und überprüft in der Mathematik letztlich deren Beweisbarkeit.

DAS KURSKONZEPT AUS SICHT DER SEMIOTISCHEN ERKENNTNISTHEORIE

Nach einleitenden Bemerkungen beginnt der Kurs mit zwei idealtypischen Problemen aus dem kaufmännischen Bereich, welche auf Systeme linearer Gleichungen führen, einem Mischungsproblem und einer Aufgabe zur innerbetrieblichen Leistungsverrechnung. Damit stehen "Mathematisierungsprozesse als Quelle von Diagrammen" (Dörfler 2006) am Anfang, wobei die Konstruktion von Diagrammen schrittweise erfolgt. Zunächst werden die Aufgabendaten in einer Tabelle übersichtlich zusammengestellt, im Fall der innerbetrieblichen Leistungsverrechnung wird die Tabelle zusätzlich in einen gewichteten Digraphen zur Erfassung des Leistungsflusses und der zugehörigen Kosten übersetzt. Diese Repräsentationen, die je für sich als Diagramme bestimmte Aspekte des Problems abbilden, vermitteln zwischen der Sachsituation und dem eigentlichen Mathematisierungsziel, dem zu erstellenden Gleichungssystem, das ein mathematisches Modell der Situation darstellt und als eigenständiges Objekt weiter untersucht wird.

Das Herstellen der Normalgestalt basiert auf einer ersten zu beobachtenden Gesetzmäßigkeit, die in eben dem Begriff "Normalgestalt" ihren allgemeinen Ausdruck findet. Der Hinweis auf die tatsächlich mögliche Größe von Gleichungssystemen in realen Anwendungen belegt die Notwendigkeit eines schematischen Lösungsverfahrens. Dieses wird anknüpfend an das aus der Mittelstufe bekannte Additionsverfahren als Gauß-Algorithmus entwickelt und an geeignet gewählten Beispielen verdeutlicht. Das Verfahren lässt sich allgemein als systematisches Operieren auf dem gegebenen Diagramm beschreiben. Seine wesentliche Idee, die Herstellung einer Dreiecksgestalt, findet ihren exemplarisch auf 4×4-Systeme bezogenen Ausdruck in einer neuen Inskription, die sinnfällig und doch nur auf der bisher erarbeiteten Wissensgrundlage verständlich ist:

$$\begin{array}{lclclcl}
\leftarrow\leftarrow\leftarrow\leftarrow = \leftarrow & & \leftarrow\leftarrow\leftarrow\leftarrow = \leftarrow & & \leftarrow\leftarrow\leftarrow\leftarrow = \leftarrow & & \leftarrow\leftarrow\leftarrow\leftarrow = \leftarrow \\
\leftarrow\leftarrow\leftarrow\leftarrow = \leftarrow & \longrightarrow & \Upsilon\leftarrow\leftarrow\leftarrow = \leftarrow & \longrightarrow & \Upsilon\leftarrow\leftarrow\leftarrow = \leftarrow & \longrightarrow & \Upsilon\leftarrow\leftarrow\leftarrow = \leftarrow \\
\leftarrow\leftarrow\leftarrow\leftarrow = \leftarrow & & \Upsilon\leftarrow\leftarrow\leftarrow = \leftarrow & & \Upsilon\,\Upsilon\leftarrow\leftarrow = \leftarrow & & \Upsilon\,\Upsilon\leftarrow\leftarrow = \leftarrow \\
\leftarrow\leftarrow\leftarrow\leftarrow = \leftarrow & & \Upsilon\leftarrow\leftarrow\leftarrow = \leftarrow & & \Upsilon\,\Upsilon\leftarrow\leftarrow = \leftarrow & & \Upsilon\,\Upsilon\,\Upsilon\leftarrow = \leftarrow
\end{array}$$

Die Lösung eines linearen Gleichungssystems in Dreiecksform durch Rückeinsetzen wird als eine fortgesetzte, wenn auch verschleierte Anwendung des Gauß-Algorithmus ausgewiesen. Hierfür ist eine geeignete Interpretation der vorhandenen Darstellungsmittel erforderlich. Der ebenfalls exemplarisch für 4×4-Systeme geführte Nachweis, dass die Schritte des Gauß-Algorithmus Äquivalenzumformungen sind, vollzieht die erste umfangreichere Deduktion von Konsequenzen aus dem gewonnenen Verfahren. Der Beweis ist für den Insider elementar, enthält aber für Schülerinnen und Schüler ggf. neu zu erlernenden Gepflogenheiten beim Umgang mit Beweisen und basiert damit auf kollateralem Wissen der mathematischen Community, in das die Lernenden erst hineinwachsen müssen. Lernen kann insofern als progressive Teilnahme an einer sozialen Praxis des Gebrauchs von Diagrammen und des diagrammatischen Schließens angesehen werden (Dörfler 2006).

Da der Gauß-Algorithmus nur mit den Koeffizienten eines Gleichungssystems arbeitet, empfiehlt es sich, lediglich das Zahlenschema der Koeffizienten zu notieren. Durch diese ökonomische Erwägung ist eine Modifikation des Darstellungssystems, nämlich der Übergang zur Matrix-Vektor-Schreibweise nahe gelegt. Sie stellt, wie sich im Verlauf des Kurses zeigen wird, in stenographischer Notation zwei universell einsetzende Sprachmittel zur Verfügung: Vektoren als n-stellige Listen und Matrizen als (m,n)-stellige Zahlenschemata. Jedoch sind wegen dieser stenographischen Kürze ähnlich wie bei Zahldarstellungen im Stellenwertsystem zusätzliche Regeln zu beachten: Die Reihenfolge der Unbekannten muss von vornherein festgelegt und dann festgehalten werden, und Nullen sind stets hinzuschreiben.

In diesem veränderten Darstellungssystem lässt sich die alte Schreibweise für ein lineares Gleichungssystem mit Koeffizienten und Variablen als Matrix-Vektor-“Multiplikation” nach bestimmten Regeln herstellen. Durch diese neue Verknüpfung werden Matrizen und Vektoren zu eigenständigen Rechenobjekten. Die abkürzende Schreibweise

$$A\vec{x} = \vec{b} \qquad (*)$$

unterstützt den Objektcharakter und erreicht eine "neue diagrammatische Schicht" (Dörfler 2003), die eigene Möglichkeiten eröffnet:

- Mit den gewonnenen Sprachmitteln lässt sich das Lösen eines linearen Gleichungssystems zu im Kontext einer binären Operation interpretieren. Es bedeutet nichts anderes, als die Lösung der Matrix-Vektor-Gleichung (*) anzugeben. Ein Vektor $\vec{z}$ ist genau dann eine Lösung der Gleichung, wenn das Matrix-Vektor-Produkt von A mit $\vec{z}$ den Lösungsvektor $\vec{b}$ liefert.
- Zur Entfaltung der Implikationen des Darstellungssystems gehört es, spezielle Matrizen und Vektoren (Einheitsmatrix, Nullmatrix, i-ter Einheitsvektor) zu betrachten, ihr Verhalten in der Matrix-Vektor-Multiplikation zu erforschen und entsprechende Regeln zu formulieren. Das Experimentieren mit der Matrix-Vektor-Multiplikation führt auch auf die Beobachtung neuer Phänomene wie die Existenz von Nullteilern.
- Die neue Operation kann auch in Bezug auf Gemeinsamkeiten und Unterschiede zum Rechnen mit (reellen) Zahlen betrachtet werden und vertiefte Einsicht in die üblichen Rechengesetze von $\mathbb{R}$ ermöglichen.
- Schließlich können die für die Matrix-Vektor-Multiplikation festgestellten Beziehungen "in einer Form von Protokollen" (Dörfler 2003) mit Hilfe von Formeln, die wieder neue Diagramme darstellen, festgehalten werden, so z. B. die Linearitätseigenschaften

  $$A(\vec{x}+\vec{y}) = A\vec{x} + A\vec{y}, \quad A(r\vec{x}) = r(A\vec{x}),$$

  aus denen wiederum spezielle Eigenschaften für homogene Gleichungssysteme hergeleitet werden können.

Die zugehörigen Übungsteile enthalten wie die Erarbeitungsteile wesentliche Tätigkeiten, die zum diagrammatischen Denken gehören (zu der folgenden Aufstellung s. Dörfler 2006):

- Das Manipulieren von Diagrammen nach gegebenen Regeln, z. B. die numerische und algebraische Ausführung von Matrix-Vektor-Operationen und das Lösen von konkret gegebenen Gleichungssystemen nach dem Gauß-Algorithmus. Diese dienen dem Vertrautwerden mit den Operationen und ihren internen Eigenschaften, sie vermitteln technisch-handwerkliche Versiertheit im Umgang mit diesen und unterstützen das Erkennen von Gesetzmäßigkeiten und Beziehungen. Unter dem Aspekt der Binnendifferenzierung im Unterricht sollte diese Tätigkeitsebene von allen Schülerinnen und Schülern beherrscht werden.

- Das Experimentieren mit Diagrammen und das Erforschen ihrer Eigenschaften, z. B: Matrizen mit bestimmten Eigenschaften suchen, Folgerungen aus den Linearitätseigenschaften ziehen und diese bei Berechnungen vorteilhaft nutzen, die Wirkung einer Diagonalmatrix bei Multiplikation mit einem Vektor untersuchen und in Worten beschreiben.
- Untersuchen von Beziehungen zwischen verschiedenen Typen von Diagrammen, z. B. das Übersetzen zwischen Tabellen oder Graphen einerseits und Gleichungssystemen andererseits bei der Lösung von Anwendungsproblemen.
- Erfinden und Entwerfen von Diagrammen, z. B. beim Aufstellen von Gleichungssystemen zur Modellierung von Anwendungssituationen.

RÉSUMÉE

Die Analyse sollte zeigen, in welch differenzierter und zugleich handfester Weise der untersuchte Kurs die Schülerinnen und Schüler in Prozesse der mathematischen Wissensbildung einführt.

LITERATUR

Artmann, B. & Törner; G. (1980): Lineare Algebra. Grund- und Leistungskurs. Göttingen: Vandenhoeck & Ruprecht.

Artmann, B. & Törner, G. (1984): Lineare Algebra und Geometrie. Grund- und Leistungskurs. 2., neubearb. Aufl. Göttingen: Vandenhoeck & Ruprecht

Dörfler, W. (2003): Diagrammatisches Denken in der Linearen Algebra. Henn, H.-W. (Hrsg.): Beiträge zum Mathematikunterricht 2003. Hildesheim, Berlin: Franzbecker, 189–192.

Dörfler, W. (2006): Diagramme und Mathematikunterricht. Journal für Mathematik-Didaktik 27, H. 3/4, 200–219.

Hoffmann, M. H. G. (2001): Skizze einer semiotischen Theorie des Lernens. Journal für Mathematik-Didaktik 22, H. 3/4, 231–251.

Hoffmann, M. H. G. (2005): Erkenntnisentwicklung. Ein semiotisch-pragmatischer Ansatz. Frankfurt am Main: Vittorio Klostermann.

Hoffmann, M. H. G. (2006): Semiotik in der Mathematikdidaktik. Journal für Mathematik-Didaktik 27, H. 3/4, 171–179.

Strang, G. (1976): Linear Algebra and its Applications. New York: Academic Press.

Törner, G. (1982): Erfahrungen und Bemerkungen zu Kursen in Linearer Algebra. Der mathematische und naturwissenschaftliche Unterricht 35, H. 6, 321–325.

CHAPTER 8

MATHEMATIZING DEFINITIONS OF BELIEFS

Norma Presmeg
Illinois State University

A Review of Törner, G. (2002). Mathematical beliefs—a search for a common ground: Some theoretical considerations on structuring beliefs, some research questions, and some phenomenological observations. In G. C. Leder, E. Pehkonen, & G. Törner, (Eds.), *Beliefs: A hidden variable in mathematics education?* (pp. 73–94). Dordrecht, The Netherlands: Kluwer Academic Publishers.

With the incisive logic of the mathematician he is, Günter Törner examined definitions of *belief* in the current literature on the role of beliefs in the teaching and learning of mathematics, noted the lack of consistency in usage and focus, and set out to structure and systematize key aspects of these definitions. In doing so, he accomplished more than a systematization of the literature in this significant area of influence in mathematics education. Characteristically, he introduced his own mathematical symbolism in this analysis, and in this way attempted to mathematize a phenomenon of human experience—that of *believing*—that some may consider to be a far cry from the certainty and predictability of the logic of mathematical principles. What did he accomplish in this mathematization? How may the results of his work be used to enhance understanding of the role of beliefs of vari-

The Montana Mathematics Enthusiast, pages 93–97

ous kinds in mathematics education? Does his model provide a theoretical framework for further research on this significant topic? These and related questions are addressed in the following review of Günter Törner's accomplishments in his chapter in the book on *Beliefs: A hidden variable in mathematics education?* that he edited with Gilah Leder and Erkki Pehkonin.

INTRODUCING MATHEMATICAL TERMINOLOGY AND SYMBOLISM IN AN ATTEMPT TO SYSTEMATIZE DEFINITIONS OF BELIEFS

The style of writing in this chapter is succinct, yet lucid and rational. Working out the core elements of a definition of beliefs is a major goal, as summarized in Törner's own words as follows.

> Specifically, a four-component definition of beliefs is presented. The model focuses on belief object, range and content of mental associations, activation level or strength of each association, and some associated evaluation maps. This framework is not empirically derived but is based on common characteristics of the literature on didactics, particularly mathematics didactics. This effort towards achieving a precise definition can provide new understandings of fundamental issues in research on mathematical beliefs and give rise to new research questions. In particular, it allows description of the term "belief systems" allowing clustering of individual beliefs into a system across each of the four components. Furthermore, it makes sense to distinguish between global beliefs, domain-specific beliefs and subject-matter beliefs. (p. 73)

In each of the four components of the definition, mathematical symbolism is introduced. This symbolism is surely a means of keeping track of the complexity of the human phenomenon of *believing*, but it is also a possible means of deepening the issues that are the foci of attention in research in this field. Before introducing each component, Törner builds a rationale for his decisions and stance, and in the process demonstrates an impressive familiarity with relevant literature. For instance, he notes that previous work has not adequately distinguished between knowledge and beliefs, but he does not pursue this avenue because "beliefs can be viewed as at the periphery of knowledge (Ryan, 1984)," as he later elaborates in his model.

Quoting Fischbein, Törner notes that "Beliefs are not mere residuals of more primitive forms of reasoning. They are genuinely productive, active ingredients of every type of reasoning" (p. 75). Further, beliefs filter the processes used in establishing the "virtual realities" in our mental processing of sense data. The profound conclusion of this observation is that in the subjective nature of this processing we do not all see the same phenomena, even when observing the same things: *We see what we believe!* The implications of this insight for mathematics teaching and learning are tremendous. However, Törner does not pursue this aspect in the chapter.

The four components of a belief, **B**, are as follows.

1. There is a belief object, **O**, because it is always a belief *about something* (p. 78). Further, there is a breadth dimension, a ***range of associations***, which makes up the content set, $\mathbf{C_O}$, related to the object O. Some of the elements of the set C_O may be conflicting (p. 79).
2. Because C_O is a *fuzzy set*, some elements of the set are central and others more peripheral. Thus it is necessary to specify the *range and content of mental associations* by means of a ***membership degree function***, μ_i, whose values vary from 0 to 1 (p. 79), i.e., $0 \leq \mu_i \leq 1$. One advantage of this notation is that it incorporates the distinction of beliefs from knowledge. Knowledge would consist in a set C_O in which the μ_i functions are equal to 1 for all values of i.
3. The ***activation level or strength of each association*** may also be invoked by special membership degree functions, μ_i, with values again varying from 0 to 1. These functions may measure
   a. levels of certitude;
   b. levels of consciousness; and
   c. levels of activation (pp. 80-81).
4. Affective evaluations of beliefs take place (as implied in many definitions). Thus one or more ***evaluation map(s)*** ε_j are defined for the range of a belief C_O and with a linguistic value scale that may be bipolar or continuous, e.g., degrees of liking or disliking (pp. 81–82).

It is also necessary to keep in mind that there is a ***person P***, who has professed the belief or to whom the belief is attributed, and that this phenomenon took place at a certain ***time t of constitution*** (p. 82). For a person P at a time t, Törner summarizes the situation as follows.

> In short, a belief constitutes itself by a quadruple $B = (O, C_O, \mu_i, \varepsilon_j)$, where O is the debatable belief object, C_O is the content set of mental associations (what traditionally is called a belief), μ_i is the membership degree function(s) of the belief, and ε_j is the evaluation map(s). (p. 82)

Although the foregoing definitions may be precise in their mathematical terminology, they are not closed, because they take into account the complexity of human variations in experience, at various times, within certain contexts, and above all they allow for the variability of human thinking and reacting to experiences. The formulation also allows for the clustering of beliefs according to the four domains in the definition, resulting in *belief systems*, and the tripartite further distinction of beliefs that are global, domain-specific, or subject-matter specific (pp. 85-87). Various hierarchies are possible in this organization. Törner perceives that "in many contexts it is not sufficient to study beliefs; the analysis of belief systems must take priority" (p. 84).

IMPLICATIONS FOR RESEARCH ON BELIEFS IN THE TEACHING AND LEARNING OF MATHEMATICS

The foregoing gives some indication of what Törner accomplished in his mathematization of a definition of beliefs. The possibilities are now open for the results of his work to be used to enhance the teaching and learning of mathematics. Such enhancement is likely to come through the avenue of research that uses his lucid categories as theoretical lenses with which to define research parameters and analyze research results, in the investigation of beliefs of various kinds in mathematics education.

Although the main thrust of this chapter is theoretical, in a final broad section before his conclusions, Törner offers "some qualitative observations" (p. 88) with regard to beliefs and their role in attitude changes. In this section he also gives readers an illustrative glimpse into his own research on mathematical beliefs. In particular, six preservice upper-secondary school teachers were asked to write freely about their experiences with calculus lessons in response to the prompts, "How I experienced Calculus at school and university", "How I would have liked to have learned Calculus", and "How I would like to teach Calculus" (p. 88). Analysis of the data yielded seven types of beliefs related to the teaching and learning of calculus. In this illustrative account of the study, two students' beliefs regarding the role of logic in calculus are contrasted: for one it is a barrier to be overcome, for the other it is something far more foundational, involving elements of epistemology. One insight as a result of this investigation is that it is necessary to consider domain-specific beliefs (such as those expressed by these preservice teachers) in terms of global views of the nature of mathematics itself. In research on mathematical beliefs, ontology and epistemology are intertwined.

Törner does not indicate how the lenses of his mathematized definition of beliefs helped the analysis in the research reported. The use of his framework in research is left open. It may be that the purpose of the mathematization of the definitions was not to provide a re-usable theoretical model for research, but for the precise clarification of the components themselves. However, what he has accomplished may be gauged in comparison with some results in the history of mathematics that he introduced earlier in the chapter (p. 77), where he noted that the proliferation of terms used by various authors instead of *beliefs* contributed to the lack of consistency in the literature. The very openness of the field may not be an impediment in and of itself. He pointed out that a similar situation existed with regard to beliefs about the nature of number in the 19th century: however, work on number theory continued, and Dedekind (1985/1888—mistakenly given as 1988 in the text) was moved to write about "What are numbers and what should they be?" This analysis led the way to an axiomatic definition of numbers. In a comparable way, Törner's careful and logical work to compose components of a definition of beliefs has the potential to further the field of research on the role of beliefs in mathematics education, whether or not mathematical symbolism is employed in using the components of his model. It is possible that his careful work may lead to a finer grain of understanding of some aspects of mathematics education, and in this way contribute to improving the teaching and learning of mathematics at various levels.

This is a very balanced chapter. The careful and at times profound analysis of components of beliefs and their systems was appreciated by this reviewer. Because of the rigor and clarity of his approach, Törner has provided a theoretical model that deepens our insights into the complexities entailed in processes of human believing. In doing so, he opens new possibilities and starts a new conversation in this field of considerable significance for mathematics education.

CHAPTER 9

INTEGRATING ORAL PRESENTATION INTO MATHEMATICS TEACHING AND LEARNING

An Exploratory Study with Singapore Secondary Students

Lianghuo Fan
National Institute of Education
Singapore

Shu Mei Yeo
Ministry of Education
Singapore

Abstract: This paper introduces an exploratory study on the integration of oral presentation tasks into mathematics teaching and learning in Singapore school settings. Five classes taught by different teachers in two secondary schools participated in the study. Data reported were mainly collected through classroom observations, teacher interviews, and field-notes. The study revealed that although both teachers and students encountered initial difficulties and challenges, after gaining necessary experiences and skills they can overcome these difficulties and challenges and become more positive and effective in using oral presentation in their teaching and learning. Moreover, teachers' beliefs, behavior, reaction and verbal responses to students' speech attributed to the effectiveness of using oral presentation in mathemat-

The Montana Mathematics Enthusiast, pages 99–122

ics teaching. The results suggest that, given necessary help and guidance for teachers and students, it is meaningful and feasible to integrate oral presentation tasks into mathematics teaching and learning.

INTRODUCTION

Developing students' communication skills has been widely believed to be one of the important goals in the international mathematics education community (e.g., Cai & Jakabcsin, 1996; Morgan, 1999; Sfard, 2001; National Council of Teachers of Mathematics, 1989 & 2000). In Singapore, the national mathematics syllabus explicitly states that one major aim of mathematics education in schools is to enable students to "use mathematical language to communicate ideas and arguments precisely, concisely and logically" (Ministry of Education, 2001, p. 9) and "develop the abilities ... to communicate mathematically" (Ministry of Education, 2006, p. 1). According to the syllabus, developing students' ability to illustrate, to interpret, to explain, and to discuss mathematical ideas and their experiences in learning mathematics has its unique value in mathematics instruction, and students should be given the opportunities to speak and write in mathematics classrooms, although very often mathematics learning has been viewed to focus mainly on computational and procedural skills.

This study is part of a larger research project on integrating new assessment strategies into mathematics teaching and learning under the Centre for Research in Pedagogy and Practice, National Institute of Education, funded by the Ministry of Education, Singapore. The main research introduces four alternative assessment strategies into daily learning and teaching in Singapore's primary and secondary mathematics classrooms. These four alternative assessment strategies are based on the use of communication assessment tasks, performance assessment tasks, project tasks, and student self-assessment tasks. The study presented herein reports the results from two secondary schools whereby teachers and students used oral presentation as one form of the communication assessment tasks.

For the larger project, we classified communication tasks into two types: written communication tasks and oral communication tasks. More specifically, the larger study employed the use of journal writing tasks as written tasks and oral presentation tasks as oral tasks. Although journal writing and oral presentation are both not prevalent in the context of Singapore schools, results from some small-scale structured research works using written and oral communication tasks in the mathematics classrooms (e.g. Yeo, 2001; Yazilah & Fan, 2002; Seto, 2002) seem to suggest the increasing recognition and interest in the use of these alternative strategies. However, available research document in this line is also found to have most focused on students' written communication skills. Research documentation, particularly classroom-based, of how mathematics teachers can efficiently engage students in meaningful learning with focus on students' oral communication skills is lacking (e.g., see Fan et al., 2006).

RESEARCH OBJECTIVES AND QUESTIONS

This study aims to investigate how oral presentation can be effectively integrated into classroom teaching and learning in Singapore school settings. More specifically, there are three main research questions:

1. What are the impacts of using oral presentation tasks on teachers' teaching methods and behaviors?
2. What are students' general perceptions and attitudes toward the use of oral presentation tasks in their mathematics learning?
3. How oral presentation tasks can be effectively integrated into the process of classroom teaching and learning?

By addressing the above three research questions, we hope to provide research-based evidences concerning the implementation of oral presentation tasks in Singapore secondary mathematics classroom settings, explore practical ways for teachers to effectively integrate oral presentation tasks into students' learning of mathematics, and offer useful suggestions for teachers to develop students' oral communication skills.

THEORETICAL FRAMEWORK AND PERSPECTIVES

The theoretical background for this study is, to a degree, drawn on three major theorists that are identified to have highlighted the crucial role that students' activity plays in students' learning: socio-cultural, constructivist and multiple intelligences.

In brief, the socio-cultural perspective gives the priority to social and cultural processes over individual thought processes and attempts to reform classroom practice by promoting less hierarchical, more interactive, and more networked forms of communication within the classroom (Goos, Galbraith, & Renshaw, 1999). Constructivist theorists believe that a learner's knowledge is constructed by the learner, and social discourse is one of the powerful ways students come to change or reinforce conceptions and knowledge construction. Student-to-student dialogue or teacher-to-student discussion are ways whereby a communicative culture can be fostered, encouraging effective meaningful sharing of information and learning (Brooks & Brooks, 1993, pp. 101–118). In addition, multiple intelligence theory claims that different students have different learning styles, strengths and experiences, and not all students learn in the same way (also see Adams, 2000), therefore students should be provided with a multitude of learning opportunities, including verbal communication opportunities.

This study views the use of oral presentation tasks from two perspectives. First, it is regarded as an alternative mode of assessment for teachers to gather information about their students' learning of mathematics and hence make relevant instructional decisions. Second, it is also viewed as a tool for developing students' communication skills. While there is no one formal definition of what oral presentation is, it is evident in the literature that oral presentation is an activity of sharing ideas and clarifying understanding verbally. One general purpose of oral presentation in mathematics classroom is to allow teacher to hear what students are thinking about mathematics, and how they express mathematics and their understanding of mathematics in their own words. Furthermore, according to the *Communication Standard for Grades 6–8* by the National Council of Teachers of Mathematics (NCTM), teachers using oral presentation tasks must provide opportunity for students to think through questions and problems; express their ideas; demonstrate and explain what they have learnt; justify their own opinion; and reflect on their own understanding and on the ideas of others (NCTM, 2000, p. 272).

We generally classified oral presentation tasks into two categories: pre-structured oral presentation tasks and impromptu oral presentation tasks that are not pre-planned but impromptu (i.e. tasks are carried out without being planned earlier or rehearsed). The following gives a brief description of what impromptu and pre-structured oral presentation tasks are:

Impromptu oral presentation tasks	Pre-structured oral presentation tasks
• Specific tasks: • Questions that are posed during instruction; • Students' responses to questions that are posed or asked; • Students' work that are represented on the board during instruction; • Students' work that are given in the homework, class work, worksheets, or textbooks; • Students' summary of a lesson.	• Specific tasks: • Questions that are pre-designed and given to students prior to lessons; • Students' previous writing tasks on their learning reflection or perceptions; • Students' previous solutions to test questions; • Students' previous writing tasks about mathematics; • A selected topic that is pre-agreed before discussion; • To report results or findings of a project work; • To report pair or group work discussion.

To illustrate how a pre-structured oral presentation task could look like, take for example the topic of Hire Purchase, a teacher could ask students to bring in newspaper cuttings and advertisement about purchase of items that involved both cash and hire purchase scheme. The questions that teacher wants students to think and work on before the presentation may look like this:

(a) Discuss and write down that is agreed among yourselves in the group the understanding about hire purchase.
(b) Find an advertisement which has both cash and hire purchase scheme. Cut out and paste the advertisement on the worksheet.
(c) Suppose you are to advice your parents about purchasing the product (that you have cut out earlier in part (b)), what would you say to them and what advice would you give?

In another example, a teacher may want to collate the common mistakes that students have made in a test or examination about the expansion or factorization in algebraic expressions. The worksheet should contain the work examples of students that have made mistakes in their test/examination, a column for students to work out the correct solutions, and a column for students to write out how mistakes pertaining to the specific questions could be prevented. After some time of either individual or small group work, students can therefore present and discuss about their work and ideas.

An impromptu oral task, for example, is that a teacher could elect a student and ask the student to summarize the day's lesson. Other students could help to 'fill-in' the gap the student may have missed out. In this way, others would also be given the opportunity to voice their opinions and ideas although they are not the one to give the summary.

It should be emphasized that in this study, our focus was on the integration of oral presentation tasks into the daily classroom teaching and learning, therefore it is our perspective that the topics of presentation chosen should closely follow and support the current instructional content and syllabus.

RESEARCH METHOD

Schools

Two secondary schools were randomly selected to participate in the study. One secondary school was identified as a high-performing school because it was selected from the 50 best performing secondary schools that offered both Express and Normal Academic streams,[1] based on the year

1999–2002 GCE "O" Level Examination results released by the Ministry of Education (MOE), Singapore. The other secondary school was identified as a non-high performing school because it was not in this top 50 school ranking. The main reason for us to select a high-performing and a non-high performing school was to observe the possible similarities and differences among different types of schools about the use of oral presentations concerning the research questions described earlier.

Classes and Students

In each school, one first year secondary (Grade 7) class in Express Course and one in Normal (Academic) Course, each taught by a different teacher, were selected to participate in the study. The study needed to follow up the students for a period of about 18 months, that is, students were tracked as they moved from Secondary One level (Grade 7) to Secondary Two level (Grade 8). However for the second year schooling, the students from the non-high performing school were streamed at the end of their first year. As a result, some students were lost due to streaming. In addition, due to some unforeseen reasons, one teacher in the non-high performing school was replaced by another teacher who was also teaching another parallel class; therefore that class was also included in the study but only the students who were involved in the first year were tracked and data analysis included only these students. Nevertheless, the study was able to track slightly more than 50% of the students.

All the classes had about 40 students, a standard size of all Singapore secondary classes. The participating teachers had a variety of teaching experiences in terms of the number of years they had taught. All the classes were observed and all the teachers involved were interviewed for data collection.

Data Collection and Analysis

There were mainly four instruments designed for data collection in this study: oral presentation tasks, students' questionnaire surveys, structured classroom observations, and interviews with teachers. Field-notes were also taken during the study.

As mentioned earlier, in the larger research project, both journal writing tasks and oral presentation tasks were introduced into the mathematics classrooms. However, due to the fact that both of them were new tasks for teachers as well as students, therefore, for teachers and students to better manage the relatively new strategies, only journal writing tasks were used in the classrooms during the first nine months of the intervention period. Oral presentation tasks were mainly introduced in the second year of the study as an integral part of teaching and learning process for a total period of two school terms (about six months). The oral tasks were first designed mainly by the researchers, then gradually by the teachers with the help of the researchers, based on the theoretical framework and perspectives discussed above. The tasks were also designed based on teachers' scheme of work. On average, four to five oral presentation tasks were done in each of the experimental classes.

Structured classroom observation was intended to document the instructional practices occurring in the classes. During the period of about 18 months, each class was observed for an average of 12 times, and each observation period was about 60 minutes. Field-notes were taken in all the classroom observations. At times, researcher was also in the classroom to video tape the teachers implementing oral presentation tasks. One of the aims was to provide teachers with feedback, advices and guidance about the use of the new tasks, as well as to communicate with the teachers and students constantly. The other reason was to gain more ideas from the classroom observations so that subsequent tasks could be constructed or designed.

In the larger research project, experimental students were asked to take a pre-questionnaire survey just before the start of the main study, and a post-questionnaire survey after the end of the intervention period. Both the questionnaire surveys were intended to measure students' attitudes toward mathematics and their learning of mathematics. For this study, we will focus on the last 16 question items in the post-questionnaire survey that were constructed to measure students' general feelings, beliefs and perception about their own ability to perform oral presentation tasks. To respond to these items, students had to agree or disagree on a nine-point Likert-type scale format with the following anchors: *1 = Disagree totally, 2 = disagree a lot, 3 = Disagree, 4 = Disagree a little, 5 = Neither disagree nor agree, 6 = Agree a little, 7 = Agree, 8 = Agree a lot and 9 = Agree totally.*

Interviews were conducted with all the five participating teachers. The interviews focused on their experiences in carrying out oral presentation tasks during teaching, the difficulties and challenges they faced, the specific actions they took to help their students, and the measures they developed to improve the integration of the task more efficiently. Each interview took about 30 minutes and was audio-recorded and then transcribed.

Qualitative method was mainly employed to analyze the data which were collected from the classroom observations, students' questionnaire surveys, and the teacher interviews, with the purpose of documenting the teacher's instructional behavior and providing general descriptions about the integration of oral presentation tasks during mathematics teaching.

It should be pointed out that this study is exploratory in nature. The study was designed to follow up with students and teachers for a period of about 18 months. A limitation is, however, during this long period of intervention, there were certain variables which were difficult to predict and/ or control. For example, in one school, the administration had decided to stream the Secondary One students at the end of the year. Thus, students in both the experimental and other parallel classes were changed at the beginning of the second year of the study, i.e., when students began their academic year of Secondary Two. As a result, students who had received or not received intervention right at the beginning of the study were all mixed up. Thus, a number of students were 'lost' in the experimental classes. Nevertheless, the study managed to track about 50% of the original students from the beginning of the study in that school. In another case, the teacher in the experimental Express class was not able to follow up with the experimental class because she had to go on a long term maternity leave, and therefore the teacher had to be replaced. Hence, the changes of administration policy in school regarding the issue of streaming and teacher transfer were some variables that were beyond the control for this research.

Another limitation of the study was related to participating students' attitudes and behavior. Although gathering evidence from students' performing on oral presentation tasks was intended to be an alternative way of assessing students' learning and understanding of mathematics, but throughout the whole intervention period, none of the teachers' evaluation on students' work on these new assessment tasks was taken into account as part of students' final school achievement grades. This arrangement is understandable because participating students should not be disadvantaged and disconnected from the schools' normal assessment practices; otherwise, these students would be negatively affected for their grades and reports. However, researchers believed that this could lead to some students not seriously attempting the oral presentation tasks, especially the pre-structured ones. Moreover, the participating teachers were observed to have tried to make time from their usual pre-planned teaching schedule to do the new tasks and this often led to teachers doing the new tasks only when they were reminded or told, which as one can see might have generated some negative influences on the results of the study.

RESULTS AND DISCUSSIONS

Below are the main findings obtained from the data we collected, in accordance with the three aforementioned research questions.

What are students' general perceptions and attitudes toward the use of oral presentation task in their mathematics learning?

As mentioned earlier, a total of 16 questionnaire items were constructed to measure three aspects of students' general attitudes toward the use of oral presentation tasks: (a) perceptions about their own ability to perform, (b) beliefs in the usefulness, and (c) acceptance of oral presentation tasks. We shall now first report results from the high-performing school, followed by results from the non-high performing school.

The data reveled that students in the experimental Express Course responded positively to only four items pertaining to oral presentation. Table 9.1 showed that although students did not generally have positive acceptance towards the use of oral presentation (Q38 & Q51), students had positive beliefs about the usefulness of doing oral presentation. For instance in Q42, 45% of the students believed that *'doing mathematics oral presentation helps me to be more aware of my understanding of mathematics* (vs. 32.5% who disagreed), 42.5% of the students believed that '*doing mathematics oral presentation makes me think broader and deeper about mathematics*' for Q47 (vs. 35% who disagreed), and more than 70% of the students also had the same opinion that '*listening to other classmates' oral presentation is helpful for me in learning mathematics*' (Q49).

Interestingly, the data for the experimental Normal Academic students also revealed somehow similar results as to the experimental Express students' responses to these 16 items. Table 9.2 showed that the Normal Academic students did not show very positive responses to items corresponding to their acceptance towards the use of oral presentation. However, the data did reveal that most students generally believed in the usefulness of oral presentation. For instance, more than 60% of the students 'agreed' that *'listening to other classmates' oral presentation is helpful for me in learning mathematics'*.

In the non-high performing school, somehow similar results to those from the high-performing school were also observed. Table 9.3 revealed that the experimental Express students had responded quite negatively about their acceptance towards the use of oral presentation in the mathematics classroom. For instance, only 33.4% and 14.4% of the students 'agreed' to item Q38 '*I like to do mathematics oral presentation during mathematics lessons*' and item Q51 '*I would like to have more mathematics oral presentations for my mathematics lessons*' respectively. However, data revealed that more than 60% of the students believed that '*doing mathematics oral presentation is not a waste of time*' (Q53) and that '*oral presentation skill is important in mathematics learning*' (Q43). In addition, there were high percentages of students who believed that doing oral presentation helped in their learning of mathemat-

TABLE 9.1 Percentages of experimental Express students (from high-performing school) responding to the 16 items in the questionnaire survey

	Disagree totally	Disagree a lot	Disagree	Disagree a little	Neither disagree nor agree	Agree a little	Agree	Agree a lot	Agree totally
General beliefs about oral presentation									
Q40: Doing mathematics oral presentation helps me to learn mathematics.	10.0%	7.5%	10.0%	15.0%	17.5%	15.0%	15.0%	5.0%	5.0%
Q42: Doing mathematics oral presentation helps me to be more aware of my understanding of mathematics.	7.5%	5.0%	10.0%	10.0%	22.5%	20.0%	20.0%	2.5%	2.5%
Q43: Oral presentation skill is important in mathematics learning.	12.5%	7.5%	12.5%	12.5%	22.5%	10.0%	17.5%	.0%	5.0%
Q47: Doing mathematics oral presentation makes me think broader and deeper about mathematics.	2.5%	2.5%	15.0%	15.0%	22.5%	17.5%	17.5%	2.5%	5.0%
Q49: Listening to other classmates' oral presentation is helpful for me in learning mathematics.	2.5%	.0%	2.5%	10.0%	12.5%	20.0%	32.5%	7.5%	12.5%
Q52: Doing mathematics oral presentation makes me learn mathematics better.	7.5%	15.0%	7.5%	5.0%	20.0%	20.0%	17.5%	5.0%	2.5%
Q53: Doing mathematics oral presentation is a waste of time.	10.0%	7.5%	7.5%	2.5%	37.5%	5.0%	12.5%	5.0%	12.5%

Q44: I am able to express about my feeling through mathematics oral presentation.	7.5%	5.0%	20.0%	15.0%	20.0%	12.5%	12.5%	5.0%	2.5%
Q45: I am able to tell others about my understanding of mathematics through mathematics oral presentation.	7.5%	.0%	20.0%	10.0%	17.5%	20.0%	17.5%	2.5%	5.0%
Perceptions about own ability to do oral presentation									
Q39: Doing mathematics oral presentation is easy to me.	15.0%	10.0%	17.5%	15.0%	20.0%	12.5%	5.0%	5.0%	.0%
Q41: I am not afraid of doing mathematics oral presentation.	15.0%	2.5%	15.0%	17.5%	12.5%	7.5%	25.0%	5.0%	.0%
Q46: I don't know how to get started when I am doing mathematics oral presentation.	2.5%	.0%	15.0%	7.5%	30.0%	22.5%	7.5%	5.0%	10.0%
Q48: I feel lost when I am doing mathematics oral presentation.	2.5%	5.0%	10.0%	15.0%	22.5%	20.0%	15.0%	7.5%	2.5%
Q50: I can do mathematics oral presentation well.	10.0%	7.5%	20.0%	15.0%	25.0%	12.5%	7.5%	2.5%	.0%
General acceptance towards oral presentation									
Q38: I like to do mathematics oral presentation during mathematics lessons	25.0%	12.5%	12.5%	10.0%	20.0%	7.5%	10.0%	.0%	2.5%
Q51: I would like to have more mathematics oral presentations for my mathematics lessons.	25.0%	12.5%	7.5%	12.5%	22.5%	7.5%	5.0%	2.5%	2.5%

TABLE 9.2 Percentages of Experimental Normal Academic Students (from high-performing school) Responding to the 16 Items in the Questionnaire Survey

	Disagree totally	Disagree a lot	Disagree	Disagree a little	Neither disagree nor agree	Agree a little	Agree	Agree a lot	Agree totally
General beliefs about oral presentation									
Q40: Doing mathematics oral presentation helps me to learn mathematics.	23.1%	5.1%	17.9%	2.6%	28.2%	10.3%	10.3%	2.6%	.0%
Q42: Doing mathematics oral presentation helps me to be more aware of my understanding of mathematics.	23.1%	2.6%	10.3%	7.7%	23.1%	17.9%	7.7%	5.1%	2.6%
Q43: Oral presentation skill is important in mathematics learning.	17.9%	2.6%	10.3%	7.7%	25.6%	17.9%	7.7%	.0%	10.3%
Q47: Doing mathematics oral presentation makes me think broader and deeper about mathematics.	17.9%	5.1%	7.7%	10.3%	23.1%	15.4%	12.8%	2.6%	5.1%
Q49: Listening to other classmates' oral presentation is helpful for me in learning mathematics.	12.8%	2.6%	.0%	5.1%	10.3%	30.8%	17.9%	5.1%	15.4%
Q52: Doing mathematics oral presentation makes me learn mathematics better.	25.6%	.0%	15.4%	5.1%	17.9%	20.5%	10.3%	2.6%	2.6%
Q53: Doing mathematics oral presentation is a waste of time.	10.3%	2.6%	10.3%	10.3%	23.1%	12.8%	2.6%	5.1%	23.1%

Q44: I am able to express about my feeling through mathematics oral presentation.	15.8%	5.3%	18.4%	5.3%	28.9%	10.5%	10.5%	2.6%	2.6%
Q45: I am able to tell others about my understanding of mathematics through mathematics oral presentation.	17.9%	5.1%	15.4%	5.1%	25.6%	15.4%	10.3%	2.6%	2.6%
Perceptions about own ability to do oral presentation									
Q39: Doing mathematics oral presentation is easy to me.	28.2%	10.3%	5.1%	17.9%	20.5%	10.3%	5.1%	.0%	2.6%
Q41: I am not afraid of doing mathematics oral presentation.	15.4%	5.1%	10.3%	.0%	30.8%	7.7%	15.4%	.0%	15.4%
Q46: I don't know how to get started when I am doing mathematics oral presentation.	7.7%	2.6%	12.8%	2.6%	17.9%	20.5%	10.3%	5.1%	20.5%
Q48: I feel lost when I am doing mathematics oral presentation.	10.3%	2.6%	10.3%	5.1%	20.5%	15.4%	17.9%	5.1%	12.8%
Q50: I can do mathematics oral presentation well.	25.6%	10.3%	12.8%	12.8%	17.9%	10.3%	7.7%	2.6%	.0%
General acceptance towards oral presentation									
Q38: I like to do mathematics oral presentation during mathematics lessons	30.8%	7.7%	10.3%	17.9%	17.9%	5.1%	7.7%	.0%	2.6%
Q51: I would like to have more mathematics oral presentations for my mathematics lessons.	38.5%	5.1%	10.3%	10.3%	20.5%	10.3%	5.1%	.0%	.0%

TABLE 9.3 Percentages of Experimental Express Students (from non-high performing school) Responding to the 16 Items in the Questionnaire Survey

	Disagree totally	Disagree a lot	Disagree	Disagree a little	Neither disagree nor agree	Agree a little	Agree	Agree a lot	Agree totally
General beliefs about oral presentation									
Q40: Doing mathematics oral presentation helps me to learn mathematics.	4.8%	4.8%	0.0%	9.5%	19.0%	28.6%	28.6%	0.0%	4.8%
Q42: Doing mathematics oral presentation helps me to be more aware of my understanding of mathematics.	4.8%	4.8%	0.0%	0.0%	19.0%	38.1%	19.0%	4.8%	9.5%
Q43: Oral presentation skill is important in mathematics learning.	4.8%	0.0%	9.5%	0.0%	23.8%	23.8%	23.8%	4.8%	9.5%
Q47: Doing mathematics oral presentation makes me think broader and deeper about mathematics.	9.5%	0.0%	4.8%	0.0%	28.6%	33.3%	14.3%	4.8%	4.8%
Q49: Listening to other classmates' oral presentation is helpful for me in learning mathematics.	0.0%	0.0%	0.0%	0.0%	14.3%	19.0%	42.9%	9.5%	14.3%
Q52: Doing mathematics oral presentation makes me learn mathematics better.	4.8%	9.5%	4.8%	0.0%	28.6%	19.0%	19.0%	4.8%	9.5%
Q53: Doing mathematics oral presentation is a waste of time.	28.6%	4.8%	9.5%	19.0%	19.0%	4.8%	4.8%	0.0%	9.5%

Q44: I am able to express about my feeling through mathematics oral presentation.	9.5%	0.0%	23.8%	19.0%	23.8%	14.3%	9.5%	0.0%	0.0%
Q45: I am able to tell others about my understanding of mathematics through mathematics oral presentation.	9.5%	0.0%	23.8%	4.8%	23.8%	23.8%	4.8%	9.5%	0.0%
Perceptions about own ability to do oral presentation									
Q39: Doing mathematics oral presentation is easy to me.	23.8%	14.3%	14.3%	14.3%	23.8%	4.8%	4.8%	0.0%	0.0%
Q41: I am not afraid of doing mathematics oral presentation.	19.0%	9.5%	9.5%	9.5%	33.3%	14.3%	0.0%	4.8%	0.0%
Q46: I don't know how to get started when I am doing mathematics oral presentation.	0.0%	0.0%	4.8%	4.8%	23.8%	4.8%	19.0%	23.8%	19.0%
Q48: I feel lost when I am doing mathematics oral presentation.	0.0%	4.8%	0.0%	9.5%	38.1%	9.5%	14.3%	9.5%	14.3%
Q50: I can do mathematics oral presentation well.	23.8%	14.3%	4.8%	23.8%	23.8%	0.0%	4.8%	0.0%	4.8%
General acceptance towards oral presentation									
Q38: I like to do mathematics oral presentation during mathematics lessons	23.8%	4.8%	14.3%	9.5%	14.3%	23.8%	4.8%	4.8%	0.0%
Q51: I would like to have more mathematics oral presentations for my mathematics lessons.	19.0%	9.5%	14.3%	9.5%	33.3%	4.8%	0.0%	4.8%	4.8%

TABLE 9.4 Percentages of Experimental Normal Academic Students (from non-high performing school) Responding to the 16 Items in the Questionnaire Survey

	Disagree totally	Disagree a lot	Disagree	Disagree a little	Neither disagree nor agree	Agree a little	Agree	Agree a lot	Agree totally
General beliefs about oral presentation									
Q40: Doing mathematics oral presentation helps me to learn mathematics.	16.7%	0.0%	8.3%	8.3%	25.0%	16.7%	20.8%	0.0%	4.2%
Q42: Doing mathematics oral presentation helps me to be more aware of my understanding of mathematics.	16.7%	0.0%	4.2%	4.2%	16.7%	20.8%	20.8%	8.3%	8.3%
Q43: Oral presentation skill is important in mathematics learning.	16.7%	0.0%	4.2%	8.3%	37.5%	16.7%	12.5%	4.2%	0.0%
Q47: Doing mathematics oral presentation makes me think broader and deeper about mathematics.	16.7%	0.0%	0.0%	8.3%	33.3%	12.5%	25.0%	4.2%	0.0%
Q49: Listening to other classmates' oral presentation is helpful for me in learning mathematics.	12.5%	0.0%	0.0%	0.0%	8.3%	12.5%	37.5%	8.3%	20.8%
Q52: Doing mathematics oral presentation makes me learn mathematics better.	16.7%	0.0%	0.0%	12.5%	20.8%	12.5%	29.2%	4.2%	4.2%
Q53: Doing mathematics oral presentation is a waste of time.	12.5%	8.3%	20.8%	8.3%	33.3%	4.2%	0.0%	0.0%	12.5%

Q44: I am able to express about my feeling through mathematics oral presentation.	16.7%	0.0%	8.3%	12.5%	33.3%	8.3%	16.7%	4.2%	0.0%
Q45: I am able to tell others about my understanding of mathematics through mathematics oral presentation.	16.7%	4.2%	4.2%	8.3%	25%	8.3%	20.8%	12.5%	0%
Perceptions about own ability to do oral presentation									
Q39: Doing mathematics oral presentation is easy to me.	20.8%	4.2%	20.8%	8.3%	20.8%	12.5%	4.2%	4.2%	4.2%
Q41: I am not afraid of doing mathematics oral presentation.	25.0%	4.2%	12.5%	25.0%	4.2%	12.5%	12.5%	0.0%	4.2%
Q46: I don't know how to get started when I am doing mathematics oral presentation.	8.3%	12.5%	12.5%	8.3%	25.0%	16.7%	4.2%	0.0%	12.5%
Q48: I feel lost when I am doing mathematics oral presentation.	4.2%	4.2%	16.7%	12.5%	25.0%	16.7%	0.0%	4.2%	16.7%
Q50: I can do mathematics oral presentation well.	25.0%	0.0%	12.5%	4.2%	41.7%	8.3%	0.0%	4.2%	4.2%
General acceptance towards oral presentation									
Q38: I like to do mathematics oral presentation during mathematics lessons	20.8%	4.2%	20.8%	8.3%	33.3%	4.2%	0.0%	4.2%	4.2%
Q51: I would like to have more mathematics oral presentations for my mathematics lessons.	29.2%	4.2%	20.8%	8.3%	20.8%	8.3%	4.2%	0.0%	4.2%

ics. For example, 62% and 52.3% of them 'agreed' that '*doing mathematics oral presentation helps me to learn mathematics*' (Q40) and '*learn mathematics better*' (Q52); 71.4% felt that '*doing mathematics oral presentation helps me to be more aware of my understanding of mathematics*' (Q42); 57.2% 'agreed' that '*doing mathematics oral presentation makes me think broader and deeper about mathematics*' (Q47); and over 80% 'agreed' that *'listening to other classmates' oral presentation is helpful for me in learning mathematics*' (49).

In the same sense, Table 9.4 showed that the experimental Normal Academic students also did not appear to have very positive acceptance towards oral presentation but 33.4% 'believed' (vs. 29.2% who 'disbelieved') that '*oral presentation skill is important in mathematics learning*' (Q43) and about 50% 'agreed' that '*doing mathematics oral presentation is not a waste of time*' (Q53). Furthermore, the students had a general consensus that oral presentation helped in their learning of mathematics. For examples, item Q42 '*doing mathematics oral presentation helps me to be more aware of my understanding of mathematics*' (Q42) had received 58.2% of students 'agreeing' to the item; 50.1% of the students 'agreed' that '*doing mathematics oral presentation makes me learn mathematics better*' (Q52); and 41.7% of the students 'believed' (vs. 25% who 'disbelieved') that '*doing mathematics oral presentation makes me think broader and deeper about mathematics*' (Q47).

To summarize, the results tabulated above suggest that overall students did not appear to accept more of oral presentation tasks to be done in mathematics classrooms, which is to a large degree understandable, given the fact that it is challenging as a new type of task,[2] it takes away time from their regular classroom teaching, and moreover student's work in oral presentation was not taken into account in their final grading and report, as mentioned earlier. However, the data also showed that overall students were actually rather positive in their beliefs about the benefits and usefulness of doing oral presentation task. In particular, students did by and large agree that doing oral presentation task "helps [them] to learn mathematics," "helps [them] to be more aware of [their] understanding of mathematics," and "makes [them] think broader and deeper about mathematics."

How oral presentations can be effectively integrated into teachers' mathematics teaching?

The data from classroom observations and teacher interviews revealed that, to effectively integrate oral presentation into the process of mathematics teaching and learning, both teachers and students should take various roles and responsibilities during the engagement. Under the two broad categories: pre-structured oral presentation tasks and impromptu ones, Table 9.5 provided a framework about teachers' and students' responsibilities when using these oral presentation tasks in mathematics classrooms, based

TABLE 9.5 A Framework about Teachers' and Students' Responsibilities When Using Oral Presentation Tasks in Mathematics Classrooms

When using Impromptu oral tasks:	When using Pre-structured oral tasks:
Students' responsibilities include • to answer teacher's impromptu questions; • to critique and discuss peers' solutions and questions posed; • to express, comment, explain and correct peers' work by using own ideas.	Students' responsibilities include • to express and present own ideas; • to construct knowledge with whole class and teacher; • to decide presentation format; • to design, create and invent solutions, responses, etc.
Teacher's responsibilities include • to give wait time for students' discussion; • to elicit and guide further thinking by questioning; • to check answers and reinforce correct solutions.	Teacher's responsibilities include • to help and monitor from a distance; • to facilitates discussion; • to summarizes each presentation and reinforce concepts learnt.

on the data collected from classroom observations and teacher interviews in this study.

The data also suggest that teachers' beliefs, behavior, reaction and verbal responses to students' speech also played an important role in enhancing the effectiveness of using oral presentation tasks in mathematics classrooms. For example, students reported in their feedback that more instructions and guidance should be provided, with suggestions including "train us to be more brave to speak up," "provide a guide dialog on how to present," "give us some hints when we are in doubt," "tell students and remind students in the middle of the presentation what is lacking or insufficient, and repeat what's expected," and "encourage us to elaborate our answers and speeches."

Table 9.6 summarizes a list of questioning techniques/prompts and measures that teachers can adopt for creating proper instructional environment for oral presentation activities.

In addition, the interviews conducted with teachers also revealed helpful information about how teacher can better facilitate and help improve students' performance in doing oral presentation tasks. According to the teachers, their students' ability in their command of the language (English) was a crucial factor in determining their ability to perform well in oral presentation tasks. From their daily classroom observations, teachers had identified some of the problems students had: "language problem; they can write down or use mathematical notation but not explain in English form," "they can't express themselves well, from the thought to the words portion. They do not know how to express what they don't understand," and

TABLE 9.6 Teachers' Questioning Techniques/Prompts for Oral Presentation Activities

Teachers' questions	Teachers' responding to student's speech
• Why do you think the answer makes sense (or does not make sense)? • Do you agree with the solution? You can disagree you know, no worries, just tell me what you think is not correct? • Did you all as a group (or as a pair) derived the same solution earlier? Did any one of you change your mind after discussion? Why did you change your mind? • How can you help him/her to elaborate his/her explanation? • You mentioned "____________." May you tell us why you said that? • Who/Which group has a different solution or has a different way of deriving your answers? • What do you think [the student's name] would probably have done and therefore made the mistake? • Do you have any questions for him/her/this group?	• Please elaborate further/more. • Don't worry about giving the wrong answers; just say what's on your mind. • Let's focus on what your friend has said and see if you get his/her idea. • That is a great presentation; we should encourage him/her further! • Doesn't matter if you're wrong, just say it! • Please think again! I am confident you know! • I know you can do better than this! Let's try again. • Don't just blindly do, the explanation is the most important part of this exercise. • You have done a great job! But is there anyone who thinks differently? • You are doing fine, please carry on. • It is indeed not easy to go out there and speak, but you did well! • You are doing well, just speak louder.

"they are shy and reserved; they are terrified to have to speak to the whole class, and they are afraid of making mistakes." The suggestions given from them included managing students well in class, explaining clearly the expectations to students, and providing appropriate teaching instruction. For examples, some teachers pointed out that: "teacher has to speak the right language," "give them a pleasant environment, a non-threatening environment with peers' support, an environment that they can speak freely," and "give students a positive and supportive environment. Tell clear rules and respect [students' ideas]."

It should be emphasized that most teachers believed that encouragement was a key factor to better give confidence to students who were engaging in something that was not previously the focus in their learning. For example, teachers pointed out that: "you've got to encourage them, you've got to convince them," "they need to know that it' ok to make mistakes, everyone is learning," and "it takes a change in the students' mindset, encourage them, do more often, and practice more." In addition, teachers also expressed that because doing oral presentation tasks were relatively a new experience and new emphasis in mathematics classrooms, we should not expect perfect results from both teachers and students.

The data collected from teachers' interviews also suggested that teachers felt the need to change the assessment for mathematics so as to accommodate the smooth integration of using oral presentation task in daily teaching. One teacher especially noted that in order to encourage both teachers and students to use oral presentation tasks, then one of the assessment criteria must also focus on student's performance in oral tasks. This teacher gave an example stating that although sometimes it could be very difficult to grade oral tasks but teacher's observations and statements commenting about students' competence in doing oral tasks should be taken as important information and this information should be credited as part of students' current formal mathematics assessment.

What are the impacts of using oral presentation task on teachers' teaching methods and behaviors?

It was noted in the study that the teachers encountered some initial difficulties and challenges, and needed time to get familiar to the use of oral presentation because using this type of tasks in the classroom was a new experience to them. The difficulties and challenges that the teachers faced were often the time factor, i.e., teachers felt that they did not have time to implement oral presentation task in class. This was quite understandable because most teachers felt that oral tasks were just some additional work for both their teaching and students' learning. That was to say, oral tasks were seen as add-on to the regular tasks that they had to do which were already pre-decided in their scheme of work that was planned earlier on. However, after some time with more experiences and, in this study, also help from the researchers, teachers had generally become more comfortable in using oral tasks; they did not express any extreme or apparent problem or concern. For example, teachers indicated during interviews: "After sometimes, you actually get quite familiar… know what to do." "Later on, I am clear what to look for, what to ask."

The teachers interviewed generally felt that as long as an oral task was appropriately designed to measure students' specific learning outcome, then the oral task could be used as a measurement about students' learning. These teachers also felt that some of the conventional tasks could be replaced by oral presentation tasks. For instance, one teacher pointed out that "[oral presentation tasks are] a reliable source to find out from students whether they had internalized what they had learnt rather than just solve questions. I think [a bit] less than 10% of those regular exercises should be replaced with oral tasks."

The data collected from the study also revealed that teachers made changes to their instructional procedures to cater to and address students' needs accordingly after finding out about students' understanding and learning of mathematics through the use of oral presentation activities. Initially, this was viewed as one of the difficulties faced by the teachers because class time was spent on giving students feedback. The teachers felt that it was a great challenge to make changes in their pre-planned teaching instructions as reflected in their scheme of work. However, after sometime when they gained more experiences, integrating oral tasks in their daily classroom teaching became easier.

Moreover, the results from the interview data revealed all the teachers except for the teacher teaching the Normal Academic class in the non-high performing school developed positive views towards the use of oral presentation in their daily teaching because they felt that the new task had given their students the opportunity to engage in learning mathematics in-depth. Although there was a general consensus that students who were weak in the command of the English language might be disadvantaged at the beginning, but with proper guidance and practice, they should be able to overcome this weakness. It also appears that the participating teachers had developed a better appreciation for the importance of oral communication skills in mathematics teaching and learning, and some argued that students who were weak in speaking and expressing should be given more help and chances in developing and improving themselves in this aspect in mathematics classrooms, because mathematics learning should not be seen merely as "solve and answer". Consistently, they generally expressed optimistic beliefs about the advantages of using oral presentation task on students' learning, as one teacher pointed out: "Oral task develops students holistically because you must think through on what you want to say and then how you say it; it is a higher learning skill compared to writing."

CONCLUDING REMARKS

In this paper, we introduced an exploratory study on the integration of oral presentation tasks into mathematics teaching and learning in five classes taught by different teachers in two Singapore secondary schools over a substantial period of time.

From the results obtained from this study, it appears that although both teachers and students encountered initial difficulties and challenges, with necessary experiences and skills over time they can overcome them and become quite effective in using oral presentation in their teaching and learning, therefore oral presentation can be reasonably implemented in mathematics classrooms.

The results also showed that both teachers and students overall developed positive views about the benefits and usefulness of using oral presentation tasks into their daily mathematics teaching and learning. We think these positive views are related to the nature and pedagogical values of oral presentation activities.

It was also observed in the study that, to effectively integrate oral presentation into the process of mathematics teaching and learning, both teachers and students should take various roles and responsibilities during the engagement. In particular, teachers' beliefs, behavior, reaction and verbal responses to students' speech attributed to the effectiveness of using oral presentation in mathematics teaching. Teachers need to recognize that opportunity for students to be involved in active and meaningful verbal communication is an essential process for their learning and knowledge acquisition. They also need to give students necessary guidance (including make clear the expectations) especially at the initial stage, and more importantly, create encouraging classroom environment for students to engage themselves in such communication.

Finally, given the fact that the use of oral presentation tasks is relatively new to many teachers and students, and in fact, to the general school system about teaching and assessment in Singapore classroom settings, it seems clear that timely evaluation, fine-tuning work and systemic reform in school assessment are important for the effective integration of oral presentation in teaching and learning in the long run, and this may take a long time to develop. Nevertheless, from this study we believe that given necessary help and guidance for teachers and students, it is not only meaningful but feasible to integrate oral presentation tasks into mathematics teaching and learning.

NOTES

1. In Singapore, there are four types of courses for secondary school students to take, which are Special Course, Express Course, Normal (Academic) Course, and Normal (Technical) Course. The academic achievements of students in the first two courses are generally better than those in the other two courses.
2. In relation to this aspect, the reasons commonly given by students for not enjoying an oral presentation task included "I will feel nervous and blur," "I would panic and forget what I want to say," "I feel embarrassed when other laugh," and "I feel uneasy presenting in front of many people because I have not much experience in presenting."

REFERENCES

Adams, T. L. (2000). Helping children learn mathematics through multiple intelligences and standards for school mathematics. *Childhood Education, 77*(2), 86–92.

Brooks, J. G., & Brooks, M. G. (1993). *In search of understanding: The case for constructivist classrooms.* Alexandria, VA: Association for Supervision and Curriculum Development.

Cai, J., & Jakabcsin, M. S. (1996). Assessing students' mathematical communication. *School Science and Mathematics, 96*(5), 238–246.

Fan, L., Quek K. S., Ng, J. D., Lee, P. Y., Lee, Y. P., Ng, L. E., et al. (2006). *New assessment strategies in mathematics: An annotated bibliography of alternative assessment in mathematics.* Singapore: Centre for Research in Pedagogy and Practice (CRPP), National Institute of Education, Nanyang Technological University.

Goos, M., Galbraith, P., & Renshaw, P. (1999). Establishing a community of practice in a secondary mathematics classroom. In L. Burton (Ed.), *Learning mathematics: From hierarchies to networks* (pp. 36–61). London: Falmer Press.

Ministry of Education. (2001). *Mathematics syllabus lower secondary.* Singapore: Curriculum Planning and Development Division, Ministry of Education.

Ministry of Education. (2006). *Secondary mathematics syllabus.* Singapore: Curriculum Planning and Development Division, Ministry of Education.

Morgan, C. (1989). *Writing mathematically: The discourse of investigation.* Bristol, PA: Farmer Press.

Morgan, C. (1999). Communicating mathematically. In S. Johonston-Wilder, P. Johnston-Wilder, D. Pimm, & J. Westwell (Eds.), *Learning to teach mathematics in the secondary school* (pp. 129–143). London: Routledge.

National Council of Teachers of Mathematics. (1989). *Curriculum and evaluation standards for school mathematics.* Reston, VA: Author.

National Council of Teachers of Mathematics. (2000). *Principles and standards for school mathematics.* Reston, VA: Author.

Seto, C. (2002). Oral presentation as an alternative assessment in mathematics. In D. Edge & B. H. Yeap (Eds.), *Mathematics education for a knowledge-based era* (Vol. 2, pp. 33–39). Singapore: AME.

Sfard, A. (2001). There is more to discourse than meets the ears: Looking at thinking as communicating to learn more about mathematical learning. *Educational Studies in Mathematics, 46*(1), 13–57.

Yazilah, A., & Fan, L. (2002). Exploring how to implement journal writing effectively in primary mathematics in Singapore. In D. Edge & B. H. Yeap (Eds.), *Mathematics education for a knowledge-based era* (Vol. 2, pp. 56–62). Singapore: Association of Mathematics Educators.

Yeo, S. M. (2001). *Using journal writing as an alternative assessment in junior college mathematics classrooms.* Unpublished master's degree dissertation, Nanyang Technological University.

CHAPTER 10

THE ROLE OF BELIEFS ON FUTURE TEACHER'S PROFESSIONAL KNOWLEDGE

Gabriele Kaiser, Björn Schwarz, and Sebastian Krackowitz
University of Hamburg

Abstract: In this paper qualitative results of a complimentary case study embedded in a large-scale study on future mathematics teachers' professional knowledge will be presented. Referring to the classification system discussed within the didactical scientific community we study the beliefs of future mathematics teachers about mathematics and the teaching and learning of mathematics as well as concerning future teachers' professional development. Based on interviews, the beliefs of two selected future teachers are contrasted. From this, the central role of education of pedagogical content knowledge on the development of beliefs during teacher education becomes evident.

"Doppelte Diskontinuität"—Double Discontinuity

Der junge Student sieht sich am Beginn seines Studiums vor Probleme gestellt, die ihn in keinem Punkte mehr an die Dinge erinnern, mit denen er sich auf der Schule beschäftigt hat; natürlich vergißt er daher alle diese Sachen rasch und gründlich. Tritt er aber nach Absolvierung des Studiums ins Lehramt über, so soll er plötzlich eben diese herkömmliche Elementarmathematik schulmäßig unterrichten; da er diese Aufgabe kaum selbstständig mit seiner Hochschulmathematik in Zusammenhang bringen kann, so wird er in den meisten Fällen recht bald die althergebrachte Unterrichtstradition aufnehmen,

The Montana Mathematics Enthusiast, pages 123–146

und das Hochschulstudium bleibt ihm nur eine mehr oder minder angenehme Erinnerung, die auf seinen Unterricht keinen Einfluß hat- Felix Klein[1]

At the beginning of his teacher education, a young student faces problems which did not remind him at all on what he has worked on when he was at school; of course he forgets those things quickly and completely, but, if then, after he has passed his exams, he starts working as a teacher, he shall suddenly teach exactly this elementary mathematics in a school adequate manner. As he is nearly unable to connect this elementary mathematics to his university mathematics, in most cases, he will soon come back to the old-fashioned teaching traditions and his university education remains more or less a romantic memory which has no impact on his teaching.

Dedication: *This contribution is devoted to Günter Törner on the occasion of his 60th birthday. Günter Törner has demonstrated an impressive broadness in his research work, and, besides many others, one of the main emphases of his research he has been on beliefs in mathematics and mathematics learning for which he often refers to the "Double Discontinuity" described by Felix Klein already at the beginning of the 20th century, which prevents real changes in mathematics teaching. In this paper we refer to approaches that have been developed together with Günter Törner and we hope therefore to make them fruitfully applicable in teacher education.*

INTRODUCTION

Teacher education has already been criticised for a long time without its effectiveness ever being analysed empirically on a broader base. There are only a few empirically based results about the impact of the worldwide varying teacher education systems on future teachers' knowledge and their developing of competencies (for an overview about the actual situation of empirical research see Blömeke 2004). Therefore, the International Association for the Evaluation of Educational Achievement (IEA) started an international comparative study in 2006, taking the education of mathematics teachers of the primary and lower secondary level as an example (Teacher Education and Development Study: Learning to Teach Mathematics—TEDS-M). By this study, for the first time the effectiveness of different teacher education systems were analysed worldwide. Concerning the future teachers, the aim of TEDS-M was to record data about their knowledge in three areas, in mathematics, mathematics pedagogy, general pedagogy and psychology. Beyond that, the future teachers were asked about their beliefs and their personality traits. Furthermore, institutional and curricular analyses were conducted, for which the intended and implemented curriculum of the educational institutions were recorded. The results of this study, which are being prepared by 20 countries (including USA and Germany), are expected to be out by December 2009.

Due to the problems with the conceptualisation of the theoretical foundation, on behalf of IEA a preparatory study has been conducted in six countries, the "Preparatory Teacher Education and Development Study (P-TEDS)", also participated by USA and Germany. As part of this study, at the University of Hamburg several supplementary and more detailed case studies on professional knowledge of future teachers and the influence of beliefs have been carried out. First results will be presented and discussed in this article. This supplementary study concentrates exclusively on the micro level of P-TEDS, the level of individual competence acquirement and the influence of beliefs, limited to the first phase of teacher education within a consecutive structure of teacher education.

THEORETICAL FRAMEWORK OF P-TEDS AND TEDS-M AND THE OBJECTIVES OF THE SUPPLEMENTARY CASE STUDY

The initial ideas of P-TEDS und TEDS-M are considerations about the central aspects of teachers' professional competencies as basically defined by Shulman (1986) and developed and differentiated further by Bromme (1994, 1995) and others (about discussions on theoretical professional foundation see Blömeke 2002). TEDS-M and P-TEDS are based on the conceptions of professional competencies of future mathematics teachers elaborated by Bromme and Weinert, and they are aimed at receiving data about the requirements for the professional tasks of future teachers, such as teaching and making diagnoses. Furthermore, by referring to Shulman (1986), the following three knowledge areas are distinguished:

1. Mathematical content knowledge sub-divided into:
    - The required cognitive activities of future teachers, based on fundamental ideas of mathematics such as algorithmising or modelling;
    - Mathematical content areas such as algebra or statistics;
    - Mathematical levels, i.e. school mathematics of lower secondary level or upper secondary level, school mathematics from a higher point of view and mathematics at university level;
2. Pedagogical content knowledge in mathematics sub-divided into:
    - Mathematical content areas as under point (1);
    - Teaching-related tasks of mathematics teachers like elementarisation of mathematical concepts or the diagnosis of students errors;
    - Stimulated cognitive activities of students, including amongst others problem solving or modelling in everyday life situations.

Referring to Bromme (1994) knowledge of mathematical pedagogical content knowledge (called by Bromme as subject-matter-specific pedagogical knowledge) is understood as the central field where mathematical content knowledge, general conceptions about mathematics, knowledge about curricular conceptions of mathematics teaching and aspects of teaching experiences as well as knowledge about the students perceptions interweave each other.

3. General pedagogical knowledge focussing on teaching and diagnostic questions.
4. Professional competencies comprise of affective and value-oriented aspects apart from cognitive-oriented dimensions of knowledge measured via belief components. These aspects will be differentiated according to the following belief categories:
   - beliefs about mathematics as a scientific discipline;
   - beliefs about teaching and learning mathematics;
   - beliefs about teaching at school and learning in general;
   - beliefs about teacher education and professional development.

Concerning mathematics-related *beliefs,* we refer in our complementary study to differentiations of Grigutsch, Raatz & Törner (1998) as it is done similarly by the P-TEDS-main study. They differentiate *beliefs* according to four categories that proved to be applicable for our complementary study. The above listed categories are characterised by statements about how mathematics is perceived. Basically there are:

- The formalism aspect characterises mathematics by a pure logic and formal proceeding and lingual precision. The deductive character of mathematics is put into the foreground,
- The schematic aspect which regards mathematics as the application of solving routines. Mathematics is understood as a conglomerate of special rules, formula and procedures;
- The application aspect regards mathematics as being useful in everyday life and emphasises the applicability and problem solving abilities;
- The process aspect which states among others that mathematics is felt as being intensively heuristic and as creative activity and that problem solving can be done in various individual ways by practicing mathematics.

The formalism and the schematic aspects are categorised as a static approach to mathematics and learning and teaching mathematics while the application and process aspects are categorised as dynamic approaches.

5. Personality traits in a professional and non-professional context

As these aspects cannot be considered completely within the framework of the following case study, we will not discuss them further. Within the framework of this study we do not include personality traits as well as further kinds of belief. We are focussing more on the so-called epistemological *beliefs*, subjective thoughts about the nature of mathematics and the characteristics of teaching and learning mathematics (see Törner 2002). In order to focus on the role of beliefs in teacher education, we will not give detailed descriptions on the other components of teacher knowledge. Especially differences between P-TEDS and TEDS-M are not considered, thus we refer to the framework of TEDS-M and publications in preparation on P-TEDS (see Tatto et al. 2007, Blömeke, Kaiser, Lehmann 2007). Similar conceptualisations of professional knowledge of mathematics teachers are used in other studies as well. We particularly refer to the COACTIV study, which basically deals with questions about the conceptualisation and the measurement of subject-based professional knowledge of mathematics teachers and possible correlations with students' development of achievement. (see among others Brunner et al. 2006a and the overview of Baumert, Kunter 2006). However, further conceptualisations of professional knowledge exist: Thus Ball, Hill, Bass (2005) distinguish among common knowledge of school mathematics and specialised knowledge of teaching mathematics and of mathematics acquired through professional training, but due to lack of space, we cannot go into detail here.

Within the scope of P-TEDS, we have conducted detailed in-depth case studies in order to analyse future teachers' professional knowledge and its relations to beliefs of which a few results will be described in the following. The case study is focussing on future teachers and their first phase of teacher education.

METHODICAL APPROACH

Within the framework of a number of additional—qualitative-oriented—studies to PTEDS a questionnaire with open items has been developed that concentrates on the areas 'modelling and real world context' und 'argumentation and proof' and admits more qualitative analyses. This questioning has been conducted with 80 future mathematics teachers on a voluntary base within the scope of pro-seminars and advanced seminars for future teachers at the University of Hamburg. The questionnaire consists of 7 items that are domain-overlapping designed—as so-called 'Bridging Items'. Each of the items captures several areas of knowledge and related beliefs; 3 items deal with modelling and real world examples, 3 with argumentation and proof and one is about how to handle heterogeneity when teaching mathematics. Furthermore, demographic information like number of semesters, second subject and attended seminars and teaching experiences—including extra-university teaching experiences—are collected (for first results see the supplementary study by Kaiser, Schwarz 2007 and contributions in Blömeke, Kaiser, Lehmann 2007).

Based on this questioning, 20 future teachers—participating also on a voluntary base—were asked more details by means of problem-centred guided interviews (Witzel 1985) about the above named topic areas. The guideline for the interviews contains pre-structured and open questions (ask-back questions). The interviews lasted about 45 to 90 minutes, and, in order to avoid fear of failure and feelings of assessment, future teachers instead of professors or university lecturers conducted the interviews. For the case study presented below, two contrasting interviews were chosen which proved to yield interesting results. The audio taped interviews were transcribed verbatim and then evaluated according to qualitative content analysis methods (Mayring 1997), more exactly, we referred to methods of the structuring content analysis aiming at extracting a specific structure from the material by using criteria that have been defined before. This means that by referring to definitions, to typical passages of the text functioning as so-called anchor examples as well as to encoding rules, an encoding guideline is developed to analyse and to structure the material. Our main content categories were defined in a theory-guided way and oriented towards beliefs as described above. For the category system differentiations we referred to the classification system of beliefs developed by Grigutsch, Raatz, Törner (1998) as described above in chapter 1.

Then, by means of the developed category system, the interviews and the questionnaires were structured and encoded. A generalization of the results is neither intended nor possible due to the small sample size. However, the results permit exemplary insights into possible structures of professional knowledge of future mathematics teachers but limited to the area of modelling.

CASE DESCRIPTIONS AND FIRST RESULTS OF THE STUDY

In the following two future teachers (names are changed) are introduced shortly. They are described based on the results of the interviews and the questionnaires.

Anne is studying for becoming teacher at the primary and lower secondary level in the 3rd semester. Her second subject is history. Until the time of the interview she has attended the basic lectures in mathematics, one lecture in mathematics pedagogy and several seminars in general pedagogy. She is giving private lessons regularly.

Ben is close to the final examination and aims at becoming a teacher in vocational schools. He had already gained a diploma in economics before his teacher study. His second subject besides mathematics is economics. Until present, Ben has attended numerous lectures on mathematics and mathematics pedagogy and carried out two practical trainings in school.

The Case of Anne

We start by presenting Anne's beliefs about mathematics as school subject or scientific discipline:

With Anne static as well as dynamic based epistemological beliefs can be reconstructed while static oriented beliefs are dominant. She states:

> *...also wenn ich an Mathe denke, dann ist das sehr vielfältig, halt mit Zahlen und Formeln umgehen, Gleichungen umstellen, diese ganzen Rechenarten..., knobeln, so ein bisschen was man auch in den Heften machen muss...*
>
> [... so, if I think on maths it is a many folded thing, handling numbers and formula, reorganise equations, all these operation methods, tossing, a bit of those things one had to do in the exercise book...]

To her, mathematics means abstract systems of formulae, equations and calculation rules dealing with the inversion of equations, for which one must apply formulae and know arithmetic rules. However, the fact that she also notices tossing indicates a dynamic aspect that is inhered in mathematics.

However, her epistemological beliefs are also characterised by a formalism aspect which becomes obvious when dealing with the role of proofs in mathematics and school mathematics.

> *...es gehört einfach elementar zur Mathematik dazu, so ist die gesamte Mathematik aufgebaut, dass nach und nach in den ganzen Jahren und Jahrhunderten irgend welche Sachen bewiesen wurden, so dass man sie verwenden durfte und das muss man zumindest mal gehört und gesehen haben, wie so was funktioniert. Das gehört einfach dazu.*
>
> [...it is an elementary element of mathematics, all of mathematics is structured like that, that over all the years and centuries things were proved so that one was allowed to use them and that one has at least once heard about it or seen how it functions. That's simply part of it.]

Anne is still at the beginning of her teacher education, her more traditional and static *beliefs* about mathematics are hardly surprising (see

Blömeke, 2004). However, first signs of dynamic perceptions of mathematics could be observed with Anne, possibly due to her actual experiences during her study and her practical work. Thus, as objectives of mathematics teaching she states:

> *Und auch so ein bisschen was neben Mathe, also dass Mathe eben auch für den Alltag gebraucht wird, nicht nur für den Matheunterricht.*
>
> [And a bit besides maths, that maths is also used for everyday life, not only for math lessons.]

According to this statement, for Anna, mathematics does not only mean a formal handling of numbers and equations etc. but for her mathematics is also strongly connected with everyday life. Thus, mathematics is in a way also a tool for solving everyday life questions and problems.

To resume, the majority of her statements reflect that Anna has predominantly static understanding of mathematics. Obviously, she refers to many of her own experiences when she was at school, such as doing specific formalisms (e.g. "given", "search for", "solution"), gesucht, Lösung) which regards important or rigid schematic ways of thinking when doing proofs. All in all, in her opinion formalisms (formal proofs etc.) from university study in mathematics should be integrated more into mathematics lessons at school because she expresses that her school knowledge did not provide her the knowledge she needed for her university study. These more 'traditional' *beliefs* are partly contrasting with dynamic perceptions of mathematics which emphasise a stronger relation to reality and the real world.

From her *beliefs* about learning and teaching mathematics, both perceptions of mathematics can be recognised. On the one hand we find receptive thinking of teaching and learning as can be seen from the following statement:

> *Ich mache als Zweites noch Geschichte und da ist es fast noch wichtiger als in Mathe, dass man da einen didaktischen Aufbau hat, weil man die Kinder ja nicht einfach rechnen lassen kann, ein paar Aufgaben, oder so, sondern man muss da durchgehend immer Methoden wechseln.*
>
> [As second subject I am studying history and there it is nearly more important than in maths that one has a didactical structure because one cannot let the children simply doing calculations, just some tasks, or so, but you must constantly change the methods.]

This shows that to her didactic structures of mathematics lessons are less important, because mathematics lessons mainly engage students in solving arithmetic problems.

These beliefs of Anne about learning and teaching mathematics are strongly related to static and formalism views on mathematics as described above and which are also expressed by Anne's statements about the meaning of proofs:

> *...ich finde es schon wichtig, dass sie von vornherein lernen, dass man Sachen beweisen kann, dass man es nicht nur zeigt, sondern dass da auch was Allgemeines hinter steckt.*
>
> [...I think it's quite important that they [the students] learn from the very beginning that one can prove things, that one does not only show it but that there is something universal behind it.]

The statement about proofs shows that to her teaching means strongly pure knowledge transfer which tells students how things are. However, the teacher shall not leave it at that but show that mathematics is characterised by formal proofs.

> *Für die Schule denke ich sollte man sich so langsam steigern. Am Anfang wird den Kinder ja eh nur gesagt, so ist das [...] und dann langsam dieses anschauliche, präformale Beweisen und dann sollte man aber irgendwann schon auch zum formalen Beweisen kommen. Damit die Kinder das auch mal gesehen haben.*
>
> [For school I think one should increase gradually. At the beginning children are just told that something is like that [...] and then slowly these descriptive, pre-formal proofs and then, after a while, one should come to the formal proofs. So that children have also seen it.]

This strongly receptive conception of learning is in contrast to a constructivist statement where she emphasises the necessity of teamwork for modelling tasks:

> *...also nach dem Beispiel hab ich schon gesehen, dass es eben Schüler gibt, die das völlig falsch angehen könnten. Und darauf hab ich gedacht, das ist immer ganz günstig, wenn sie einen Partner haben, mit dem sie sich absprechen können. [...] Einfach als Hilfestellung, dass sich auch keiner allein gelassen fühlt, weil er gar nicht weiß wie er es machen soll.*
>
> [...After looking at that example, I have seen that there are students who start with it in a totally wrong way. And after that I have thought that it is always profitable if they have a partner with whom they can discuss. [...] Just as assistance so that nobody feels left alone because he does not know how to do it.]

It becomes clear that she prefers active learning and teamwork in groups to individual learning. In this context it is important to let all students participate in the learning process and to support the own solution attempts of the students.

Und dass man die Kinder auch dahingehend unterstützt, dass man sagt, ja das ist ein toller Ansatz und mach das mal weiter. [...] wenn man die Kinder vorher schon lobt, oder wenn sie auch wissen, dass sie was Richtiges machen, dann trauen sie sich auch.

[And that one helps children that way that one says ‚that's a good approach, just go on". [...] if one compliments children for something or if they also know that they are doing something correctly, they are more encouraged.]

This ambivalence between the procedure of prescribing solutions and approaches and encouraging to do something in one's own way is a red thread running through the whole interview as the following statement shows exemplarily. There she describes how she would act, if learning difficulties occur:

...oder dass ich einfach Mal etwas vorgebe, dass ich sage, rechnet doch das mal aus und dass sie dann verstehen, [...]. Und dass sie dann vielleicht auf die Idee kommen das weiter zu führen.

[...or that I just give something in advance, that I tell them to calculate it so that then they will understand [...]. And that then they will get the idea to continue with it.]

Her beliefs about mathematics teacher education and professional education of future mathematics teacher strongly emphasise the necessity of a solid subject-related education as essential basis of teacher education:

...das Fachwissen ist, denke ich, einfach unabdingbar. Es muss auf jeden Fall da sein, denn wenn ich das Fachwissen nicht hab, kann ich auch die Didaktik nicht lernen, das ist so ein bisschen von einander abhängig...

[...the subject-related knowledge I think is indispensable. It must exist, because if I don't have that subject-related knowledge I also cannot learn the didactics, they depend a bit on each other.]

However, basically, she argues that at primary and lower secondary level the emphasis is put on the children while at the upper secondary level on science.

Ich möchte halt wirklich mit Menschen arbeiten und ich hab mich nur schwer getan auf Oberstufe zu studieren, allein der Mathematik wegen und hab mich dann aber für die Kinder entschieden und deswegen mache ich Mathe jetzt für das GruMi-Lehramt.

[I really want to work with people and I had difficulties in studying for the upper secondary level, especially because of mathematics and then I decided for the children and therefore now I am studying for teaching at the primary and lower secondary level.]

This corresponds with her emphasis on empathy, humanity, patience and diagnostic competence teachers should have. Teachers should be able to put themselves into the students' position and to recognise their problems, to show the needed feeling and sympathy and they must be persistent in giving explanation.

> *Viel Geduld, auf jeden Fall mathematisches Verständnis, aber eben auch dieses Einfühlsame, dass sie verstehen, wer verweigert, wer kann einfach nicht und dann ein bisschen auch das Gefühl dafür wer kann vielleicht mehr, wo kann man noch reizen. Und ansonsten auch viel Menschlichkeit, glaub ich.*
>
> [Much patience, in any case mathematical understanding, but also this sensitivity for they can understand who refuses, who really is not able, and then also the feeling for who might be able to do better, or where one can demand more. And besides that, also much of humanity, I think.]

She emphasises the role of practical courses because such competences one cannot learn from theoretical education. Thus, to her practical courses and own experiences are much more important.

> *...die Geduld hab ich, glaube ich, eher bei meinen Nachhilfeschülern gelernt.*
>
> *Und die Eigenschaften kann man höchstens gesagt bekommen, was man haben sollte, aber die muss man wohl in den Praktika üben. Deswegen denke ich sind die auch schon wichtig.*
>
> [...patience, I think I have learnt more from giving private lessons.
>
> And the capacities one can only be told, which one should have, but one needs to practice them in practical courses. That is the reason, I think, they are so important.]

Altogether, her understanding is strongly influenced by her experiences with private lessons she gave to students.

For her the education in pedagogical content knowledge in mathematics forms a relevant component of the teacher education. Before that lectures she used to teach following their own feelings, now mathematics didactics provides her with a valuable tool for a qualified planning of lessons:

> *...didaktisch, denke ich, hab ich einiges gelernt, also damit habe ich mich vor dem Studium auch gar nicht mit beschäftigt [...] das habe ich bis jetzt nur aus dem Gefühl gemacht, jetzt hab ich dafür auch ein paar Regeln gekriegt.*
>
> [...didactically, I think I have learnt a lot, and before studying I have not been engaged with that [...] that I have done by my own feeling, now have also received some rules for it.]

She underlines that subject-matter pedagogical content knowledge is dependent on subject knowledge and cannot exist without it. That subject-related pedagogical content knowledge and general pedagogical knowledge make an important contribution to the development towards professionalism, she did not yet realise at the beginning of her studies. First she regarded those seminars just as gap fillers in her timetable between subject-related courses before she realised its benefit:

> *Ja, auf jeden Fall sehr wichtig, weil ich festgestellt hab, dass ich vieles jetzt einfach nur so mitgenommen hab, weil es steht ja auf meinem Stundenplan. Und man dann erst im Nachhinein erkennt, okay, dafür könnt ich das jetzt noch gebrauchen…*
>
> [Yes, by all means, it is important because I have learned that many things I have just chosen because it is on my timetable. And that only afterwards one realises, okay, I can need it for that too.]

To resume, it can be stated that in her opinion subject-based education is important especially regarding the own profession, because the pedagogical content knowledge in mathematics depends on solid mathematical knowledge. Furthermore, meanwhile pedagogical content knowledge in mathematics has gained greater importance because it became fundamentally useful for choosing and justifying methods when planning lessons. In connection with the development of professionalism she gives a central role to practical experiences, especially concerning the development of personality traits and the communication with students.

The Case of Ben

In Ben's beliefs about mathematics as a subject or a scientific discipline, dynamic perceptions are dominant. For Ben mathematics starts from questions and problems of everyday life and our natural environment for which he strongly separates himself from a perception which values mathematics as an abstract and formalistic structure.

> *Das hat was mit deinem Leben zu tun und Mathematik umgibt uns. Es geht hier gar nicht um Taylorreihen oder irgendetwas Abgefahrenes, sondern um ganz praktische einfache Sachen.*
>
> *…Mathematik ist für mich keine abgehobene, theoretische Wissenschaft für abgehobene, theoretische Wissenschaftler, sondern etwas, was mit meinem Leben zu tun hat und zwar immer dann, wenn da Zahlen und die Verbindung von Zahlen oder Mengen eine Rolle spielen.*

[It has something to do with your life and mathematics surrounds us. It is not about taylor series or spacy things but about purely practical and simple things.

...for me, mathematics is not a theoretical aloof science for theoretic aloof scientist, but something which has to do with life, and always has to do with life when numbers or the connections between numbers or amounts are playing a role.]

Relation to reality and everyday life is a very important part of his *beliefs* about mathematics. He understands mathematics as an instrument which helps him to handle everyday life problems better.

Dann brauche ich unter Umständen mathematisches Handwerkszeug, um Sachen vergleichen zu können, zusammenfassen zu können, auch bewerten zu können, wenn ich irgendwo eine Statistik lese und mir da erzählt wird, da ist irgendwo ein Anstieg von zehn Prozent von irgendetwas, dann muss gerne wissen, was bedeutet das.

[Then I possibly need a mathematical instrument which enables me to compare things, to resume, to evaluate, when somewhere I read a statistic that tells me that there is an increase of ten percent of something, then one will be happy if one knows what that means.]

Formal aspects only play a minor role for him: He underlines that formal proofs are also meaningful but they are much less meaningful than the solution of everyday problems by the application of mathematics.

Ich habe selbst in meinem Matheunterricht [...] viel mehr Techniken und Prozeduren gelernt, als inhaltlich verstanden, worum es geht. Ich habe erst hier an der Uni erfahren, wofür Analysis und Ableitungen da sind. [...] Wozu ist das eigentlich gut, im richtigen Leben, in der Praxis. War mir nicht klar.

[I myself have learned more techniques and procedures in my mathematics lessons [...] than I have understood the contents of them and what they where about. I have learned not until the university what's the use of calculus and derivates. [...] For what one can use that, in real life, in practice. That was not clear to me.]

For him, the real world context is an indispensable part of mathematics. He emphasises the modelling circle which plays a central role when dealing with real world examples:

...einmal die Reduzierung der Daten, die mir die Realität vorgibt auf das mathematische Modell, da muss ich filtern und mich auf das, was ich wesentlich finde beschränken. Wenn ich da grobe Fehler mache, dann ist natürlich meine mathematische Lösung zwar aus dem mathematischen Modell richtig errechnet, aber sie sagt mir nichts oder wenig über die Realität, oder möglicherweise sagt sie mir etwas Falsches.

[on the one hand the reduction of the data, which was given to me on the mathematical model by reality, there I have to filter and concentrate on what I consider to be essential. If I make a gross error there, than of course my mathematical solution is indeed calculated correctly out of the mathematical model, but does not tell me anything or few about reality, or potentially tells me something wrong.]

Within the modelling process he considers the phases of mathematisation and interpretation as especially important, the formal calculation is less important to him. For him the work within the mathematical model is a formalism, which cannot be executed without prior mathematisation und which is more or less worthless without reinterpretation. Again one can see clearly that for Ben using mathematics as a tool for solving extra-mathematical problems is more important than calculating mathematical models according to a fixed scheme.

Corresponding with this, one can find his statements about the significance of proving in mathematic lessons, which he considers to be important, as proving makes the generality of certain issues clear for the students. On the other hand he wants to choose the kind of proof highly in relation to the level of the learners:

Wenn ich sehe, das Niveau der meisten Schüler ist so, dass sie alle mit dem präformalen Beweis glücklich sind, dass aber die meisten von denen den formalen Beweis noch vertagen könnten, dann sollen sie ihn auch gerne haben. Wenn ich das Gefühl hätte oder sogar die Gewissheit, dass ich bei 90 Prozent der Schüler, wenn ich jetzt den formalen Beweis mache, die Erkenntnis aus dem Präformalen wieder zerstöre, dann würde ich sagen lassen wir den.

[If I see, the level of most of the students is like they are all happy with the preformal proof, but most of them could do with the formal proof, then they can willingly have it. If I had the feeling or even the certainty, that I would destroy the cognition out of the preformal for 90 percent of the students, if I did the formal proof now, then I would say we leave it out.]

His *beliefs* about teaching and learning mathematics correspond with his dynamic conceptions of mathematics as a process or activity. Thereby he describes the high algorithmic orientation and its focus on learning and executing schematic procedures as a problem of German mathematics lessons.

...ich habe auch in der Hospitation mal einen Lehrer gesehen, der bei Dreisatzaufgaben immer schwer nach Schema gegangen ist [...] ich würde viel mehr darauf drängen, einmal das Prinzip zu verstehen und sich eben inhaltlich klar zu machen, bei so einer Aufgabe [worum es geht]...

> [...once I saw a teacher during one of my school practicum, who always proceeded with proportional exercises strictly on the basis of a scheme [...] I instead would insist much more on the one hand on understanding the principle and on making clear for oneself, such an exercise [what it is all about]....]

He in contrast highlights that it is important to show the practical use of mathematics for the students´ everyday living and by doing so to motivate the students and to take away their mathematics anxiety.

> *Wichtig ist natürlich auch den Schülern die Angst zu nehmen [...] auch mit dem Hinweis darauf, das kann nicht nur der Eine, der hier gut in Mathe ist, sondern das könnt ihr alle. Ihr müsst hier keine Angst vor haben, es beißt nicht und ihr könnt heut einfach nach Hause gehen und sagen, das habe ich gemacht, das weiß ich, wie das geht und da habe ich was verstanden.*
>
> [Of course it is also important to take away the anxiety from the students [...] also with hint that not only the one who is good in maths here can do it, but all of you can do it. You don´t have to have any fear of it here, it doesn´t bite and you can simply go home today and say, I have done this, I know this, how it works and there I have understood something.]

Thereby practical, action-orientated aspects should be in the foreground also for proving, mathematics should be made concrete and descriptive, and that is why pre-formal proofs should play a decisive role in teaching mathematics. Though his concept of mathematics is a constructivist and students–orientated one, responding to the strengths and weaknesses of the learners and trying to produce references in their language and to tie up to their conceptions. Doing so for him mutual respect plays a central role, which means an engagement on the part of the teacher who takes the students serious and is really interested in them. These students–orientated conceptions of teaching mathematics are accompanied by a high reference to group working in teaching in which the teacher is more a moderator instead of giving the impulses and where the students clarify their understanding problems on their own first.

His beliefs towards teacher education and towards the professional development of future mathematics teachers are influenced by the experience of the distance between his mathematical education at the university and his practical experiences in school. In Ben's opinion his studies of mathematics at the university have prepared him meanly for the mathematics needed in school and in his opinion this part of his studies is more or less a necessary leg which one has to do for becoming a teacher but which has no use for his professional qualification. The motivation which is necessary to "struggle through" theses studies only results out of his practical experiences in school.

> *Gerade Mathe an der Uni, unzählige Tage, an denen ich dachte ich bin hier falsch, ich bin zu doof, was auch immer, aber wenn ich damit an der Schule war, dachte ich, doch dafür lohnt es sich durch das Studium zu kommen.*
>
> [Especially maths at the university, countless days when I thought I am wrong here, I am too stupid, whatever, but when I was at a school with this, I thought, however that is worth coming through the studies.]

In his opinion there is little use in the studies of mathematics for the later subject teaching, in contrary one has to feed on one's own school experiences.

> *Das heißt, wenn mich jetzt jemand fragt, sagt mal schnell Geometrie neunte Klasse, kann ich mir das Buch nehmen und möglicherweise kann ich mir das auch relativ schnell zusammenreimen, aber zusammenreimen heißt aus eigener Schulerfahrung, nicht aus dem Studium.*
>
> [That means, if someone asks me now, say geometry ninth grade quickly, I can take the book and maybe I can make sense of it relatively quickly, but making sense of it means by own school experiences not by the studies.]

Furthermore he stresses the necessity to be taught a broad knowledge basing on school mathematics, quasi a "sweeping statement" for mathematics, giving you a general overview and preparing you for your later work as a teacher. He attributes a high significance to the didactical education for the professional development and thinks that problems in teaching mathematics at school often result out of a lack in knowledge related to this area.

> *...ich muss keine hohe Mathematik machen, sonder wirklich relativ einfache Geschichten. Das höchste der Kunst ist, dass ich in der höheren Handelsschule mal ganz bisschen Kurvendiskussion und ganz bisschen Matrizenrechnung mache. [...] Und da sehe ich wirklich die Herausforderung mehr in der Didaktik...*
>
> [...I don't have to do high mathematics, but truly relatively simple stories. The highest of the art is that I sometimes do very little curve sketching and very little matrices arithmetic at the commercial college. [...] Und there I really see the challenge more in the didactics.]

He stresses the necessity of a higher significance of the pedagogical content knowledge in mathematics which has a too low value especially in contrast to the mathematical studies which are provided by the university with a higher amount of contact hours per semester and a higher acceptation. Hence in his opinion the didactics are often completed secondarily and sometimes are used as "gap fillers".

> *Man hat im Grundstudium Sorge vor der Zwischenprüfung Analysis vor allem. Aber, wenn man die gemacht hat, da guckt man wann im Stundenplan Platz ist und klatscht*

dann irgendwie diese Fachdidaktik noch mit rein und das ist dann schnell erledigt, macht keinen Stress. Man muss halt vielleicht da sein, aber so nach dem Motto: Da kann man auch gar nicht durchfallen...

[One worries about the intermediate examination, calculus mainly during the basic studies. But, when you have finished it, then you look when there is space in the schedule and then you slap these subject didactics also in it somehow and that is done quickly then, cause no stress. Maybe you have to be there, but themed like: There you can´t fail.]

Besides the didactical training he also emphasizes the high value of personal characteristics and a general interest in children and the work with them as important for a profession as a teacher.

...die Liebe zur Person, zum Menschen, dass ist etwas, was den Pädagogen auszeichnet. Ich muss ein Interesse daran haben, an den Schülern, mit denen ich da zusammen arbeite. Das sind nicht Schüler Nr. 1 bis 25, sondern die haben Namen, die haben alle eine eigene Biographie.

[...the love towards a person, towards human beings, that is something, what is characterizing the pedagogue. I have to have an interest on it, on the students, with whom I am working with. These are not my students no. 1 to 25, but they have names, the all have there own biography.]

In summary his statements about the education at the university clearly show that for him as a future teacher at a vocational school especially the education in mathematics as a science has little value for his later work. He points out that he has not been able yet to use mathematical contents learned at the university for his own teaching but always had to go back to the knowledge of his own schooldays during his practical phases. He stresses the necessity of a higher importance of lectures in pedagogical content knowledge in mathematics which in his opinion are taught in a much too small extend. On the other hand in his opinion one will acquire a big amount of one's professionalism during the second and third phase of the teacher education and during one's own teaching, provided that one has the required personal characteristics for becoming a teacher which can not be taught by any training's activities.

SUMMARY AND CONCLUSIONS

The results of the analyses of the interviews about *mathematical beliefs* can be summarized on the following three levels

- Concerning the *beliefs* towards mathematics as a subject one can find the biggest differences between Anna as a student in her third semester and Ben as a student being shortly confronted with his examination. Anna shows a more or less static view on mathematics whereas dynamical conceptions clearly dominate in Ben's opinion.
- With regard to the *beliefs* towards teaching and learning Ben's dynamical view continues in constructivist concepts about teaching and learning mathematics. Ben understands mathematical lessons as an active process starting with questions and problems resulting from everyday life. One the one hand similar *beliefs* can be reconstructed from Anna's statements. But on the other hand she also often shows beliefs emphasising the more receptive side of teaching and learning mathematics. This corresponds with her more static view on mathematics.
- Concerning the professional development of teachers both Anna and Ben stress the high importance of practical experiences and subject based didactical education.

In detail:

With regard to the *beliefs* toward mathematics as a subject a more statistic view on mathematics, that is a view basing on mathematical formalisms, appears in Anne's answers as she is just at the beginning of her studies. This view could nearly be expected because Anne is a beginner in university study and starts her studies influenced by her own experiences of mathematics lessons in school. Because of these experiences she tends to traditional opinions towards teaching and learning mathematics, a fact which is indicated on in literature time and again (cf. Ball, Lubienski und Mewborn, 2001). Her beliefs are impressed by a static tendency because obviously her own mathematics lessons were mostly algorithmic and schematic orientated, less dynamical and related to reality.

Ben shortly confronted with his examination disposes of several years of practical and professional experience. With him a clear predominance of dynamical *beliefs* towards the subject mathematics appears, which means more process- and application-orientated beliefs focussing more on an everyday life relation and less on executing algorithm. This is in accord with the state of the discussion about changes of attitudes during one' studies which point out that students change from conservative ideas of education at the beginning of their studies to more liberal attitudes at the end of their studies (cf. Dann et al. 1978). Furthermore this positive influence of the studies on the development of the *beliefs* can also be observed with Anne already to some extent, as she not only has a traditional formal view on mathematics, but also liberal attitudes towards mathematics can be reconstructed.

For both, Anna and Ben, their respective *beliefs* towards teaching and learning mathematics are strongly influenced by their previous experiences. Thus a tightened attitude towards mathematics and its educational value, which is also reflected in his *beliefs* towards the way how teaching works, appears with Ben at the end of his studies. His experiences from his previous completed vocational training and from the following studies of mathematics obviously have strongly influenced his *beliefs* towards teaching and learning mathematics. So one can see in his statements the closeness to constructivist concepts of practicing teachers as pointed out by the research done by Leuchter et al (2006) on the comparison of German and Swiss teachers. For example he phrases decidedly that mathematical lessons have to tie up to students' life and to their concepts to be successful. Mathematical lessons should be fewer teachers centred but more focussing on the active doing of mathematics and on active learning processes of the students. Furthermore one can reconstruct dynamical, application-orientated beliefs towards mathematical lessons, which shows that for him the extra mathematical use and the understanding of the contents are the focal points of mathematical lessons in contrary to a algorithm orientated use of techniques and procedures.

One can clearly identify the "filter function" of beliefs in Ben's ideas. Pajares (1992) describes this function as follows:

> Thought processes may well be precursors to and creators of beliefs, but the filtering effect of belief structures ultimately screens, redefines, distorts, or reshapes subsequent thinking and information processing. (p. 325)

Thus Ben's view on mathematics influences one the one hand strongly his concepts about teaching and learning mathematics, and on the other hand his ideas about the knowledge necessary for his future professional work.

In contrary more still "swinging" *beliefs* towards teaching and learning mathematics can be reconstructed with Anne as a beginner with her studies. She partly has liberal conceptions about teaching, such as the use of open teaching methods or the avoidance of compulsion towards students. But on the other hand her *beliefs* towards teaching and learning often are still conservatively influenced describing the teacher as a pure transmitter of knowledge. Also her rather schematic conception of mathematics, with training schematic algorithm being in the foreground, corresponds with these rather conservative ideas. Here Anna obviously is strongly influenced by her own school days, and she has not yet gained enough experiences from practical courses in school which could have offered her different perspectives. Contrariwise Anna sometimes confines herself consciously from her own school experiences, which she - to some extent - does not want to adopt into her own teaching practice. She expects from her entire studies to learn ways to do it "better" in the future.

Relating to the *beliefs* towards education of mathematics teachers and their professional development, both Anne and Ben see a positive influence of the studies—or at least parts of them - on the professional development of mathematics teachers. Anna and Ben attribute a high importance to the pedagogical part of their studies and there especially to the pedagogical content knowledge in mathematics. Both describe that they could sensibly capitalise on their courses in mathematics didactics, because there they were given concrete tools for planning lessons, which they only had developed on a gut level before. Especially Ben highlights that the didactics are neglected at the university und do not take the position which is entitled to them. He attributes a significant role for teachers' professionalizing to education in pedagogical content knowledge in mathematics.

These corresponding statements indicate on the big potential of didactical courses in view of changing beliefs, as these didactical courses have a direct relation to the further professional field und hence can directly take up with the beliefs related to mathematical teaching. However, both Anne and Ben regard the practical experiences, which will be made during and after their university studies as the most important part of teacher education and teachers professionalizing. These self conducted lessons and furthermore the practical work with children and students allow an active learning process in which beliefs and cognition can be gained or changed by own experiences.

In contrast differences concerning the valuation of the studies appear in relation to the time one has already been studying. Anna as a beginner still has high expectations towards her teacher education at university. She expects a big benefit for her own professional development from the entire university studies, which means also from the pure mathematical education. Ben as a student in the end of his studies is considerable more "disenchanted". He sees little use for his later practice in school especially in the knowledge taught in mathematics. For him teacher education at the university fails the needs of the later demands in practice and so does not help with regard to the professionalizing of becoming a teacher.

Contrasting both students and their beliefs in general one can reconstruct the first part of the so called "Bowl of Konstanz" ("Konstanzer Wanne") as described in literature. According to the research of a research group from Konstanz about the change of attitudes of future teachers, these future teachers have quite "conservative" beliefs at the beginning of their studies which decline during their studies in favour of more "progressive" beliefs. After first contacts with real work in school practice these progressive beliefs are—in consequence of the so called practical shock—substituted again by more conservative beliefs being geared to conformism with the teaching staff and tending to more compulsion towards the students (cf. Dann et al. 1978). This so called "Bowl of Konstanz" can be demonstrated roughly as follows (the arrows mark the beginning of the studies and the practical shock):

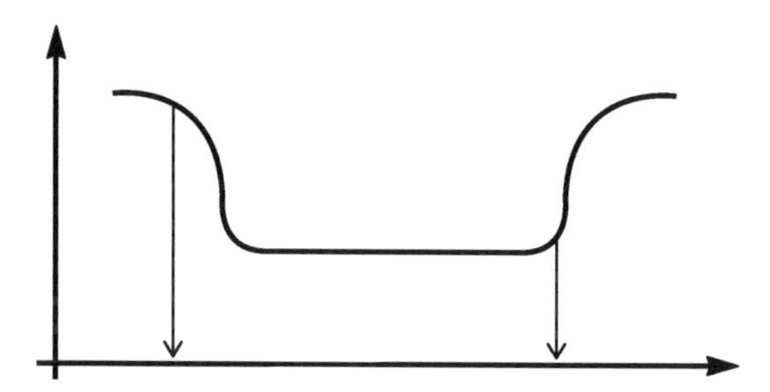

Although these descriptions are not without controversy, they show a clear closeness to the Double Discontinuity described at the beginning by Felix Klein, who traces this development to subject reasons. Precisely the description of the practical shock and as a consequence of it the return to traditional teaching methods, according to Felix Klein, is also released by the circumstance that the university has not provided the knowledge necessary for practical work in school. This criticism is formulated exactly in this way by Ben who is very disappointed by his studies, especially by the education in mathematics and who wishes more knowledge focusing on school knowledge. This demand correspond with the conceptualisation of pedagogical content knowledge developed in PTEDS and as it is described in the beginning (cf. Tatto et al. 2007, Blömeke, Kaiser, Lehmann, 2007). This demand upon teachers' professional knowledge can also be found in the work of the COACTIV-project (cf. Krauss et al., in print) and in the work of Ball, Hill, Bass (2005) who claim that mathematics teachers should have mathematical knowledge being composed of the following two components: "'common' knowledge of mathematics that any well-educated adult should have and mathematical knowledge that is 'specialized' to the work of teaching and that only teachers need know" (p. 22).

Due to the case study character of this description it is neither intended nor possible to generalise the results of these interviews. It has especially to be considered that the case study is based on a convenience sample and not a representative sample, although these two cases are based on the interviews with more students, who describe similar aspects. However, the description of the cases of Anne and Ben shows exemplarily structural weaknesses of teacher education to which Felix Klein already referred at the beginning of the twentieth century, and whose adjustment is not fulfilled. The critics is still topical, such as the remoteness from reality of teacher education, which is often complained about, the too little interlocking with practical parts of the studies, the missing relations between the taught mathematical contents and the knowledge needed in school, and the too little significance of didactical courses which could in the sense of Bromme (1994, 1995) build a bridge between the different components of teacher education.

Günter Törner has worked for the change of teacher education for future mathematics teachers in many initiatives which aimed and still aim for overbearing the Double Discontinuity in the sense of Felix Klein by a well understood subject education in mathematics—that is elementary mathematics from a higher perspective. Thus, Felix Klein describes in the foreword of his "as subsuming lectures" conceived courses as its aim:

> *Meine Aufgabe hier wird stets sein, Ihnen den* gegenseitigen Zusammenhang der Fragen der Einzeldisziplinen *vorzuführen, der in den Spezialvorlesungen nicht immer genügend zur Geltung kommt, sowie insbesondere ihre* Beziehungen zu den Fragen der Schulmathematik *zu betonen. Dadurch, so hoffe ich, wird es Ihnen erleichtert werden, sich diejenige Fähigkeit anzueignen, die ich doch als eigentliches Ziel Ihres akademischen Studiums bezeichnen möchte:* daß Sie dem großen Wissensstoff, der Ihnen hier zukommt, einst in reichem Maße lebendige Anregungen für Ihren eigenen Unterricht entnehmen können.[2]

> [My task here will always be, to demonstrate to you the *mutual relation between the questions of the individual disciplines,* which not always shows to advantage in the specialised lectures sufficiently, furthermore especially to stress their *relations to questions of school mathematics.* Thereby, so I hope, will it be more easy for you, to acquire that ability, which I would like to refer as the actual aim of your studies: *that you once can take agile suggestions for you own teaching on a high degree from the vast amount of topics, which comes up to you here.*]

NOTE

1. Felix Klein (1933), Elementary mathematics from a higher perspective. Vol. 1 (*Elementarmathematik vom höheren Standpunkte aus. Erster Band.*) p. 1.
2. Ibid. pg. 2.

REFERENCES

Ball, D L., Hill, H.C., & Bass, H. (2005). Knowing mathematics for teaching. *American Educator,* Fall, 14-22, 43–46.

Ball, D. L.; Lubienski, S. T., & Mewborn, D. S. (2001). Research on Teaching Mathematics: The Unsolved Problem of Teachers' Mathematical Knowledge. In V. Richardson (Ed.), Handbook of Research on Teaching (pp. 433–456). New York: Macmillan.

Baumert, J., & Kunter, M. (2006). Stichwort: Professionelle Kompetenz von Lehrkräften. *Zeitschrift für Erziehungswissenschaft*, 6(4), 469–520.

Blömeke, S. (2002). *Universität und Lehrerausbildung*. Bad Heilbrunn: Klinkhardt.

Blömeke, S. (2004). Empirische Befunde zur Wirksamkeit der Lehrerbildung. In: S. Blömeke et al. (Eds.), *Handbuch Lehrerbildung* (pp. 59–91). Klinkhardt: Bad Heilbrunn.

Blömeke, S., Kaiser, G., & Lehmann, R. (Eds.) (to appear in 2007): *Kompetenzmessung bei angehenden Lehrerinnen und Lehrern. Ergebnisse einer empirischen Studie zum professionellen Wissen, zu den Überzeugungen und zu den Lerngelegenheiten von Mathematik-Studierenden und –Referendaren*. Münster: Waxmann.

Bromme, R. (1994). Beyond subject matter: a psychological topology of teachers' professional knowledge. In: R. Biehler et al. (Eds.): *Didactics of mathematics as a scientific discipline* (pp. 73–88). Dordrecht: Kluwer.

Bromme, R. (1995): What exactly is 'pedagogical content knowledge'?—Critical remarks regarding a fruitful research program. In S. Hopmann, & K. Riquarts. (Eds.): *Didaktik and/or Curriculum* (pp. 205–216). Kiel: IPN.

Brunner, M. et al. (2006a). Die professionelle Kompetenz von Mathematiklehrkräften. Konzeptualisierung, Erfassung und Bedeutung für den Unterricht. In M. Prenzel et al. (Eds.): *Untersuchungen zur Bildungsqualität von Schule. Abschlussbericht des DFG-Schwerpunktprogramms* (pp. 54–82). Münster: Waxmann.

Brunner, M. et al. (2006b). Welche Zusammenhänge bestehen zwischen dem fachspezifischen Professionswissen von Mathematiklehrkräften und ihrer Ausbildung sowie beruflichen Fortbildung? *Zeitschrift für Erziehungswissenschaft*, 9(4), 521–544.

Dann, H.-D. et al. (1978). Umweltbedingungen innovativer Kompetenz. Eine Längsschnittuntersuchung zur Sozialisation in Ausbildung und Beruf. Stuttgart: Klett.

Grigutsch, S., Raatz, V. & Törner, G. (1998). Einstellungen gegenüber Mathematik bei Mathematiklehrern. *Journal für Mathematik-Didaktik* 19(1), 3–45

Kaiser, G.; Schwarz, B.; & Tillert, B. (2007). Professionswissen von Lehramtsstudierenden des Fachs Mathematik. Eine Fallstudie unter besonderer Berücksichtigung von Modellierungskompetenzen. In A. Peter-Koop; A. Bikner-Ahsbahs (Eds.). *Mathematische Bildung—Mathematische Leistung. Festschrift zum 60. Geburtstag von Michael Neubrand* (pp. 337–348). Hildesheim: Franzbecker.

Klein, F. (1933, 4th edition). Elementarmathematik vom höheren Standpunkte aus. Vol. 1. Berlin: Springer.

Krauss, S. et al. (in print). Pedagogical Content Knowledge and Content Knowledge of Secondary Mathematics Teachers. *Journal of Educational Psychology*.

Leuchter, M. et al. (2006). Unterrichtsbezogene Überzeugungen und handlungsleitende Kognitionen von Lehrpersonen. *Zeitschrift für Erziehungswissenschaft*, 9(4), 562–579.

Mayring, P. (1997). *Qualitative Inhaltsanalyse: Grundlagen und Techniken*. Weinheim: Deutscher Studien Verlag.

Pajares, M. F. (1992). Teachers`Beliefs and Educational Research: Cleaning Up a Messy Construct, University of Florida. In: Review of Educational Research, 62(3), 307–332.

Shulman, L.S. (1986). Those Who Understand: Knowledge Growth in Teaching. *Educational Researcher,* 15(2), 4–14.

Tatto, M.T. et al. (2007): *IEA Teacher Education Study in Mathematics (TEDS-M). Conceptual Framework. Policy, Practice and Readiness to Teach Primary and Secondary Mathematics.* Amsterdam: IEA.

Törner, G. (2002). Epistemologische Grundüberzeugungen—verborgene Variablen beim Lehren und Lernen von Mathematik. *Der Mathematikunterricht* 48(4-5), 103–128.

Witzel, A. (1985): Das problemzentrierte Interview. In: G. Jüttemann. (Eds.): *Qualitative Forschung in der Psychologie* (pp. 227–256). Weinheim: Beltz.

CHAPTER 11

GIRLS' BELIEFS ABOUT THE LEARNING OF MATHEMATICS

Guðbjörg Pálsdóttir
Iceland University of Education

Abstract: There has been an increased attention to research on beliefs about mathematics and mathematics education and it has become one of the central elements of study in mathematics education. This paper reports from a qualitative research study on the beliefs of four Icelandic teenage girls about mathematics, the study of mathematics, and themselves as learners of mathematics. Their descriptions and thoughts are viewed in the light of theories and newfound results of the existing quantitative research on girls´ beliefs about mathematics and the study of mathematics. The main conclusions of this research are that these particular Icelandic girls:

- view mathematics as a process
- place emphasis on understanding and solving the problems at hand
- are self-confident, well organized and study hard
- do not often use elaboration strategies.

INTRODUCTION

Mathematics is one of the main subjects in primary and lower secondary school in most countries. Extensive research has been conducted in the field of mathematics education in the last thirty years. The research has

The Montana Mathematics Enthusiast, pages 147–158

contributed to the understanding of how people learn and to finding new ways of organising teaching. Research in this area has drawn people's attention towards the influence of beliefs on how people learn. I have been a lower secondary school teacher for several years and became interested in learning about beliefs and their importance. I wondered how much I knew about the beliefs of my pupils over the years and discovered that my knowledge is limited. My main effort had been to get to know my pupils as individuals and analyse their mathematical knowledge. Therefore, I was interested in deepening my knowledge on some of the already existing research on beliefs and to conduct my own research.

My main research question was:

> How do pupils in lower secondary schools in Iceland think about mathematics and their mathematical learning?

This research question was divided into three areas: (1) the beliefs about mathematics, (2) the study of mathematics, and (3) the pupils themselves as learners of mathematics. My experience had taught me that teenage girls often ask many questions about learning and their interest in mathematics is decreasing. Consequently, I decided to focus on the beliefs of only a few girls. A qualitative research methodology was used and the study was based on individual interviews with four Icelandic girls. They were all in their final year of lower secondary school.

The following definition of the concept beliefs is used in the study:

> Students' mathematics-related beliefs are the implicitly or explicitly held subjective conceptions students hold to be true about mathematics education, about themselves as mathematicians, and about the mathematics class context. These beliefs determine in close interaction with each other and with students' prior knowledge their mathematical learning and problem solving in class. (Op 't Eynde, De Corte, & Verschaffel, 2002, p. 27)

THEORETICAL BACKGROUND

Since the seventies, great interest has been shown in the study of students' beliefs and ideas. It can be expected that this research-interest is based on the attitude that beliefs influence how people understand themselves and their surroundings, and how they deal with their lives.

The main view is that on the basis of ideas on a specific matter every individual develops beliefs about it. His beliefs evolve from simple perceptual beliefs, experience, ideas, and expectations. Beliefs are built up from many factors and their interactions are complicated. Pupils develop their

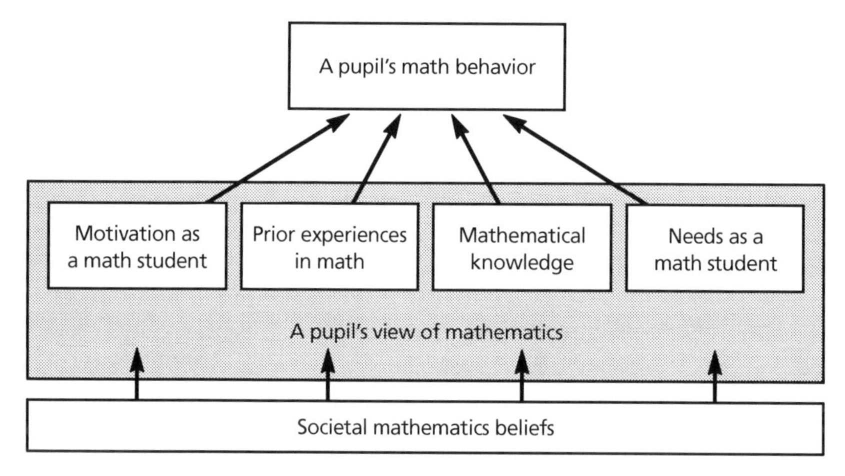

Figure 11.1 *Source:* Pehkonen & Safuanov, 1996, p. 34)

beliefs in interaction with their surroundings and they also influence their surroundings. Mathematical ideas and beliefs act as a filter that influences all their thoughts and actions concerning mathematics. Prior experience of mathematics and the learning of mathematics influence both beliefs towards learning and the use of mathematics. Societal mathematical beliefs also influence pupils' beliefs.

More factors could be mentioned that influence the mathematical behaviour of students. A network of influences from the people in one's surroundings influences the individuals' beliefs and how or if they try to learn mathematics. Beliefs towards subjects and learning are, in addition to cognitive factors, the basis of learning. Beliefs have to do with factors such as motivation, self-confidence, and how positive students are. These factors do not only support the learning, they are a part of it (Pehkonen & Safuanov, 1996; McLeod & McLeod, 2002).

Gender is one of the factors that has been found to be of great influence, but not to the same extent on the performance as on the beliefs and thereby on the motivation and the purpose of learning mathematics. Around the turn of the century some research (Brandell, Nyström, & Staberg, 2002; Pehkonen, 1994) showed that beliefs towards mathematics, the study of mathematics, and the experience of being a learner of mathematics, which were held by pupils in lower secondary schools, were changing. The pupils expressed beliefs indicating that mathematics was more for girls than for boys, and the research showed that girls worked better in maths-class and were more successful. More awareness had risen that some social factors, inside and outside the classroom, had some influence. Some research had shown that the majority of pupils no longer saw mathematics as a male-dominated subject. There was, however, a clear difference in what the sexes considered important in the learning of the subject. In the reports from PISA 2000, similar conclusions were drawn (Centre for Educational Research and Innovation, 2003, pp. 7–25, 82–90, 127–142). In PISA 2000, the boys scored higher, but the difference was not significant. The ways the sexes studied were different. The boys were more confident and showed more interest in mathematics. They believed they could cope with learning difficulties, used elaboration strategies and enjoyed competition. The girls were more concerned with what to learn and used more energy. They paid more attention to organising their study, were able to concentrate more easily and used more control strategies. The gender difference was there, but it was in beliefs and ways of learning. The conclusion was that in order to work for equality it was necessary to work with beliefs and learning methods. In Monograph 1 of the Montana Mathematics Enthusiast, the article by Steinthorsdottir & Sriraman (2007) examines the anomalies of Iceland's performance on PISA with respect to gender differences.

A big research project on teenagers' beliefs in mathematics, the GeMa-project, has been conducted in Sweden, from which Brandell and her colleagues (2002) have reported. The focus is on gender and comparison of the beliefs of girls and boys. From this, many things of interest appeared, underlining that the beliefs of teenagers are very diverse. More than half of them thought that mathematics is neither a male nor female domain. To give some ideas of how Swedish teenagers thought, I have chosen three examples:

- The girls are considered to work hard in lessons, get encouragement from the teachers, and the expectation is that they will do well.
- The boys are supposed to be disturbing in class, assumed to like using computers, like challenging problems, expect mathematics to be easy and that they will need mathematics in their future jobs.
- Girls think that it is important to understand mathematics and get worried if they are not succeeding.

These studies use quantitative methods and they gave me a good overview over this research field. Pekhonen's theories also gave me some inspiration as to what questions to ask.

METHODOLOGY

Very little research has been done on the mathematical beliefs of Icelandic teenagers. Iceland has participated in some multinational research, recently PISA 2000 (Centre for Educational Research and Innovation, 2003) and PISA 2003 (Björnsson, Halldorsson, & Olafsson, 2004). This research is entirely quantitative. I thought it would be interesting to conduct a qualitative research where I could study thoroughly the beliefs of a few individuals and give some ideas of how they express their beliefs. Many studies look at gender as a factor of great influence (Brandell, et al., 2002; Gothlin, 1999). So I decided to narrow my research and study only girls and relate my findings to the big multinational studies and some studies of girls' beliefs about mathematics. The four girls I interviewed were all 15 years old attending a lower secondary school in the capital city of Iceland, Reykjavik. They were volunteers from a class of 12 girls (and 10 boys).

There are many different research approaches in the field of qualitative research. In interviews, participants have good possibilities to use their own words and the interviewer gets real examples of how the participants express their experience and what concepts they use. In an interview, the participant also gets a chance to add new elements and the interviewer can ask new questions to get a clearer picture of the participant's ideas. I found this an interesting approach for me as a researcher entering a new field. I wanted to find out what ideas the girls had, interpret them, react and ask further. I prepared some open questions and divided the subject into three main areas, the subject, the study, and being a learner. The research question was as well divided in three main questions: What is mathematics? What is important in the study of mathematics? How does it feel to be a mathematical learner?

DESCRIPTION AND ANALYSIS

The first interview was about the girls' beliefs about mathematics. I asked how they would describe mathematics and mathematical knowledge and how they felt about it. I used in my analysis three main perspectives of the nature of mathematics, traditional, formalist, and constructivist perspective (Pehkonen & Törner, 2004). In the traditional perspective mathematics is seen as a set of skills or a toolbox. In the formalist perspective mathematics is logic and rigour and the focus is on the system. In the constructivist perspective the process in building up understanding is most important.

Helga, one of the girls, was asked the question: *What is mathematics?* She answered:

> Mathematics makes it possible for you to calculate sizes and helps you in your daily life to know what you need to know and it also tells you why things are the way they are.

In the interview Helga mainly showed the traditional perspective. She

- talked about the importance of arithmetic
- said that geometry was about using the right formulas
- thought everybody used mathematics
- saw mathematics as a big subject and thought she had still much to learn
- said that mathematics was about explaining relations

The other girls expressed similar beliefs. However, they had different views as to how they valued the usefulness of mathematics and influence in the society.

Mathematical knowledge must be the aim of the study of mathematics. Some mathematics educators have found it useful to distinguish between two types of mathematical knowledge, conceptual knowledge and procedural knowledge. In his book *Elementary and Middle School Mathematics,* Van de Walle (2004, p. 27) writes about these two main perspectives about the nature of mathematical knowledge, Conceptual Knowledge and Procedural Knowledge. Conceptual knowledge consists of logical relationships constructed internally and existing in the mind as a part of a network of ideas. Learning is based on the individual as he is building up his own knowledge. He needs to be reflective and active. He must ask his own questions and think about how he can use his knowledge to understand new things. The learning of mathematics is about making sense of how the mathematical ideas connect. The understanding will be better as the individual makes more connections. The mathematics teacher gives out problems that help the student to construct knowledge on a specific area. Discussions are important both between students and also between the student and the teacher. Elaboration strategies are in focus and students have to draw conclusions from their work.

Procedural knowledge is the knowledge of the rules and the procedures that one uses in carrying out routine mathematical tasks and also the symbolism that is used to represent mathematics. Students have to build up their knowledge step by step. Students have to be active in memorising the procedures. They use various ways to improve their memory without focusing on their mathematical understanding, as when they use rhymes to remember how to calculate fractions. The aim of procedural knowledge is to build up instrumental understanding that students can use quickly when solving routine problems. The teachers' role is to build up a logistic sequence of small pieces of mathematical information. Most important is that the teacher is able to explain and make the learning easier for the students.

In the interviews on mathematical learning the girls found it hard to talk in general terms about mathematical learning. They could nevertheless describe clearly their ideas of a good mathematics teacher and how he should play his part in the learning process.

Lara, one of the girls, said that in teaching the teacher should go from the simple to the more complicated. The teacher's role is to organise and build up a learning process. The learner's role is to follow the process and practice on many problems. She considered the role of the teacher to be very important. She thought the relationship between the learner and the teacher was meaningful. She found that the learning is most successful if the learner studies under the guidance of the teacher. Lara described a good teacher of mathematics like this:

> Mathematics teachers should take one step at a time when explaining. He should be patient and have various ways to explain the same concept. It is an advantage if the learner knows the teacher and the teacher knows the learner on personal terms.

The girls all showed a strong tendency to look at mathematical knowledge as procedural knowledge. Their view was that the learning of mathematics was important for everyone, and in order to succeed in the learning the student should work well every day and organise his work precisely. They thought that good performance involves the mastering of arithmetic skills and strategies. However, they did express, especially Soffía, that it is important that the learning advances the thinking skills and logical thinking. They also proclaimed the importance of independent students' work when building up an understanding of the procedures.

When discussing with the girls how it felt to be a mathematical learner, I used as a background some research on how people sense their own ability (Guðbjörnsdóttir, 1994; Magnúsdóttir, 2003; Fennema, 2000). This research shows that a person's estimation of one's own abilities influences how one organises and performs an act. It also has an influence on what kind of problems one deals with, how much effort one puts into it, and for how long time one tries. Further examination of the research results shows that gender influences the students' choice of challenging mathematics courses. These ideas are in coherence with the results from the PISA 2000 study (Centre for Educational Research and Innovation, 2003).

Meece discusses the studies by Fennema and Peterson (1985, as cited in Meece, 1996) in the nineties. They developed a model for learning behaviour to try to explain the gender difference in how the sexes succeeded in solving mathematical problems. They proclaim that in order to be able to solve complicated problems you need to act in a specific way. The students have to be able to get deeply involved in the problem, be able to work alone and show persistence, and concentrate. The model consists of four elements:

- Evaluation of the student's own mathematical abilities
- Sense of how useful mathematics is
- Learning abilities
- Sense that mathematics fits your gender-role (Meece, p. 117)

These are the main components of the Fennema and Peterson model. It has been used in many research studies. The fourth element in the model seems to have little influence, though the results from the other parts of the model seem to be gender biased. Fennema and Peterson claim that the difference in learning style is due to the socialisation in the classroom. Reports show that it is very common that teachers support boys in problem-solving strategies. It is also found that girls rather than boys avoid risks and problem-solving and seek guidance. This has influence on how teachers encourage their students to take part in mathematical discussions and thinking. According to Fennema and Peterson, boys are more enterprising, more active in their study, and more likely to start discussions with their teacher about what they are studying, than girls.

The interviews on how it felt to be a mathematical learner were lively. The girls' expressions were strong and personal. Their vocabulary was greater and more varied when discussing this question. They became eager in the conversation and their thoughts went deeper. Helga said:

> I find it is rather easy to understand mathematics though I sometimes need some explanations. I think it is very important that the teacher is able to explain the same thing different ways. I feel irritated if I don't understand.

Helga felt that mathematics can give both positive and negative feelings. She felt happy when she succeeded in solving difficult problems but unhappy when she failed. From her point of view mathematical learning is about solving the problems that the teacher gives.

Lara finds it important to have peace, quiet and space when she is studying because then she finds it easier to think. She claimed she was good at understanding mathematics.

> Usually I find it rather easy to understand mathematics. It is very important to understand because otherwise you will get problems later. It is also a fact that if you understand it, there is no problem to remember the strategy.

Lara emphasized the importance of understanding. She said that understanding was a precondition for her to find it interesting and challenging to do mathematics. When she was asked about the text in the mathematics books she said:

> I often read the explanations in the mathematics books but I find the strategies and methods they describe complicated. I like it much better when people show me how I am supposed to solve the problems.

Sigrun said that she found it very important to understand the things with which she was working in her mathematics study. She did not feel comfortable if she finished a problem without understanding what she was doing. The mathematics was sometimes troubling her, not least the algebra. When the learning was going badly, she felt she was "a loser" but when she understood she felt like "a genius". She connected these emotions mostly to dealing with new material. She said:

> ...always when you understand you get a good feeling, you become proud of yourself and you are confident that the rest will be easy.

Sigrun emphasized the importance of organising her study and found it easier to concentrate in peace and quiet.

> I find it most rewarding to work on projects that are challenging but not too difficult so that you can solve them after trying for a while. Then I have to use my brain and I try hard to solve problems like that.

Soffía did not find it hard to learn mathematics.

> I think I have mathematics in my genes. I understand everything, at least when I have given it a thought. It doesn't take a long time for me to learn something.

Soffía thought that in the learning of mathematics analysing what you are looking for is most important. She did not find it useful to read the text in the mathematics books or listen to the explanations from the teacher. She believed in dealing with the problems. Soffía thus described successful learning:

> I find it most rewarding to work on problems where I really have to think. I prefer problems that take me up to 30–60 minutes to solve ... and I gain most if I find the solution without any help.

The main conclusions from the girls' beliefs about themselves as mathematical learners were that these four girls felt very confident about their abilities to learn mathematics and to use them. Sigrun and Soffía expressed that it was most rewarding to struggle with the problem solving them by themselves, while Helga and Lara thought it was best to get at once an explanation if they found it difficult to understand. All the girls found it easy to organise their study and felt good when they were working on mathematical problems. All of them experienced joy when they understood things they had been working on. They all thought they were able to work independently in their study within the teacher's framework.

My main findings are that these girls express similar beliefs as girls in the overseas research-studies I used as a reference. Their beliefs about mathematics, the study of mathematics and themselves as learners of mathematics are in tandem with the conclusions of the PISA-study from 2000 (Centre for Educational Research and Innovation, 2003) and the Swedish GeMa-project (Brandell, et al., 2002).

DISCUSSION AND CONCLUSIONS

Qualitative research study is a way to give expressions of some individuals a space and a depth so that their views can be analysed from many angles. A qualitative research study based on interviews also gives the researcher opportunity to ask for the meaning of the words the individual expresses. In my research, I focused on getting a description of the girls' beliefs and not so much on how they had developed those beliefs or why. There is always a dilemma how to interpret research results and how you can use them to understand the rest of the world. In qualitative research, the main conclusion is saying something about the individuals involved, but, at the same time, being careful about using the results to draw conclusions about other people. The results can, however, often be used to understand how people think because a good insight and understanding of one individual can make it easier to understand how others think. I feel at least that my experience has made it easier for me to discuss beliefs on mathematics and mathematical learning with other teenagers and also other mathematical students. I also feel that my research has validated the idea that a research study in the Scandinavian countries and some other European countries can give some ideas of the situation in Iceland. I know that it is not possible to compare results from a qualitative research to results from quantitative research. Nevertheless, I have found it helpful to use results from a quantitative research in my analysis and to gain some knowledge in that field.

Conducting a research study is very rewarding for the researcher. I learned a lot while I was looking into these girls' minds and their ways of thinking. It would be interesting to make a follow-up study discussing those girls' beliefs at a later stage in their lives and also to use the experience from this study to interview several Icelandic teenagers.

REFERENCES

Björnsson, J., Halldorsson, A. M., & Olafsson, R. F. (2004). *Stærðfræði við lok grunnskóla.* Reykjavík: Námsmatsstofnun.

Brandell, G., Nyström, P., & Staberg, E. (2002). *Matematik i grundskolan—könsneutralt ämne eller inte?* Lund: Lund Universitet, Matematikcentrum.

Centre for Educational Research and Innovation. (2003). *Education at a glance—OECD indicators—2003 edition.* Paris: Organisation for Economic Cooperation and Development.

Fennema, E. (2000). *Gender and mathematics: What is known and what do I wish was known?* Paper presented at The Fifth Annual Forum of the National Institute for Science Education, May 22–23, 2000, Detroit, Michigan. Retrieved January 4, 2004, from http://www.wcer.wisc.edu/nise/News_Activities/Forums/Fennemapaper.htm

Fennema, E., & Peterson, P. (1985). Autonomous learning behavior: A possible explanation of gender-related difference in mathematics. In L. C. Wilkerson & C. Marrett (Eds.), *Gender influences in classroom interaction* (pp. 17–36). New York: Academic Press.

Gothlin, E. 1999. *Kön eller genus?* Göteborg: Nationella sekretariatet för genusforskning.

Guðbjörnsdóttir, G. (1994). Sjálfsmyndir og kynferði. In R. Richter & Þ. Sigurðardóttir (Eds.), *Fléttur: Rit rannsóknarstofu í kvennafræðum* (pp. 135–202). Reykjavík: Rannsóknarstofa í kvennafræðum, Háskólaútgáfan (University of Iceland Press).

Magnúsdóttir, B. R. (2003). *Orðræður um kyngervi, völd og virðingu í unglingabekk.* [Ópr. M.A. ritgerð.] Reykjavík: Háskóli Íslands (University of Iceland), Félagsvísindadeild (Faculty of Social Sciences).

McLeod, D. B., & McLeod S. H. (2002). Synthesis—Beliefs and mathematics education: Implications for learning, teaching, and research. In G. C. Leder, E. Pehkonen, & G. Törner (Eds.), *Beliefs: A hidden variable in mathematics education?* (pp. 115–127). Dordrecht: Kluwer Academic Publishers.

Meece, J. L. (1996). Gender differences in mathematics achievement. In M. Carr (Ed.), *Motivation in mathematics* (pp. 113–131). New York: Hampton Press.

Op 't Eynde, P., De Corte, E., & Verschaffel, L. (2002). Framing students' mathematics-related beliefs. A quest for conceptual clarity and a comprehensive categorization. In G. C. Leder, E. Pehkonen, & G. Törner (Eds.), *Beliefs: A hidden variable in mathematics education?* (pp. 13–37). Dordrecht: Kluwer Academic Publishers.

Pehkonen, E. (1994). Seventh-graders' experiences and wishes about mathematics teaching in Finland. *Nordisk Matematikkdidaktikk, 2*(1), 31–47.

Pehkonen, E., & Safuanov, I. (1996). Pupils' views of mathematics teaching in Finland and Tatarstan. *Nordisk Matematikkdidaktikk, 4*(4), 31–59.

Pehkonen, E., & Törner, G. (2004). Methodological considerations on investigating teachers' beliefs of mathematics and its teaching. *Nordisk Matematikkdidaktikk, 9*(1), 21–49.

Steinthorsdottir, O. &, Sriraman, B. (2007). Iceland and rural/urban girls—PISA 2003 examined from an emancipatory viewpoint. In B. Sriraman (Ed). International Perspectives on Social Justice in Mathematics Education. *The Montana Mathematics Enthusiast, Monograph 1,* pp. 169–178, University of Montana Press.

Van de Walle, J. A. (2004). *Elementary and middle school mathematics.* New York: Longman.

CHAPTER 12

MODELLING IN SCHOOL

Chances and Obstacles

Hans-Wolfgang Henn
University of Dortmund

Dedicated to Günter Törner's 60th birthday

Abstract: One goal of mathematics education is to educate pupils to become responsible citizens. Important for this is the achievement of modelling competence of students at school. I want to identify chances and obstacles which can promote the development of these competences, but can also obstruct them drastically. This will be illustrated by concrete examples.

THE RELEVANCE-PARADOX

The American fantasy writer Terry Pratchett is very popular among young people in English speaking countries as well as in Germany. One of his scenarios is a "discworld." This reminds at once of the novel Flatland by Edwin Abbott which was published in 1884 and of the many possible mathematical aspects in such a scenario. This might lead one to believe that Pratchett

The Montana Mathematics Enthusiast, pages 159–176

was also interested in mathematics. In the preface of the German edition of one of his books (Pratchett, 2005) he writes, however "I want to emphasize that this book is not in the least insane. Such a description is only apt for sappy mathematicians, who mix up geometry with the joy of living." (my translation).

Pratchett is not alone with this attitude. Mathematics, especially in German society, is seen as something which is no fun and has nothing to do with real life. My friend Mogens Niss (1994) relates this to the *Relevance Paradox*. This means that mathematics is entering deeper and deeper into more and more parts of life, but people do realize this less and less.

From my point of view, mathematics has two important sides. On one side mathematics is a special science with a special culture of thinking. Mathematics has its own aesthetics and beauty, which, however, is not accessible to everyone, same as with literature, fine arts, or music. One the other side, mathematics possesses an extraordinary functionality that allows us to bring order and understanding to all parts of our life, but also bears the danger of misuse. The aim of mathematics education needs to be to help students acquire a coherent view of mathematics. Teaching needs to be oriented towards *both* aspects and should be able to help students experience *both* of them.

Apart from elementary calculation skills the central aim of mathematics education is to acquire abilities to select appropriate methods for the application of mathematics in different areas of life. Related to this is an understanding of the phenomenon *mathematics* as cultural achievement and product of human thinking. This means that special mathematical content and details, which might be useful for students in their lives after school are of less importance compared to methods and possibilities of mathematics, because actually more than 99% of our students will not work as mathematicians in their further life.

MATHEMATICS AND THE REST OF THE WORLD

The brilliant methaphor of "Mathematics and the Rest of the World," that can be presented by the well-known picture of the modelling circuit (Fig. 12.1) originates from Henry Pollak (1979).

An application of mathematics requires a relation to the world beyond mathematics. One needs to understand something better, explain or predict certain phenomena, solve problems, provide foundations for decisions and so on. These can be a questions from science, a technical applications or a problem with a private, social or political background.

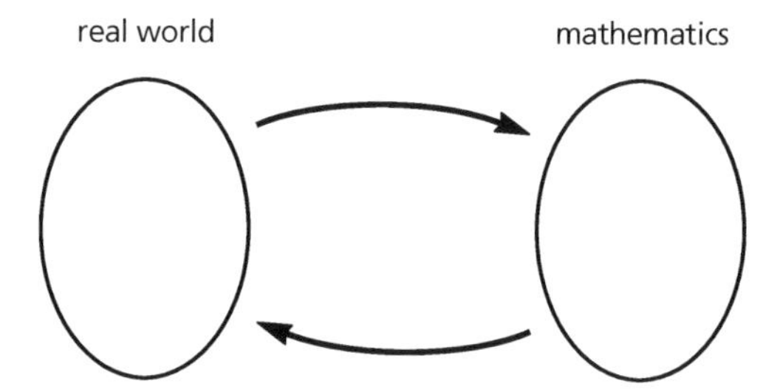

Figure 12.1 The model circuit.

Relations between reality and mathematics are especially relevant within the PISA framework (compare Neubrand, 2003). There, the construct of "mathematical literacy" was defined as

> an individual's capacity to identify and understand the role that mathematics plays in the world, to make well-founded judgements and to use and engage in mathematics, in ways that meet the needs of that individual's life as a constructive, concerned, and reflective citizen.

This concept aims directly at the future life of young people. In PISA the use of the mathematical knowledge of students is analysed within various situations and contexts of reality.

The importance of applications of mathematics and mathematical modelling is shown by the study initiated by the ICMI, the *International Commission on Mathematical Instruction,* on this topic. Recently, the Study Volume belonging to this study, *Modelling and Applications in Mathematics Education* (Blum et al., 2007), was published.

WHAT ARE MODELS?

Models are simplifying presentations, which consider only certain, somehow objectifiable parts of reality. A simple example is a map. Models are mappings from reality into mathematics. The purpose for a model is to draw conclusions for reality. Of course, for a model to be good, you must show it leads somewhere. This crucial point that models need to be helpful for something, that conclusions need to be drawn for reality and that these conclusions need to be tested in an *experimentum crucis* is often forgotten in so-called application problems in school.

Especially important are the following two aspects of models: On the one side, we have *normative models.* Examples are the law for the income tax, the methods for elections, the rules for the soccer world championship, and so on. On the other side, we have *descriptive models* which we differentiate be-

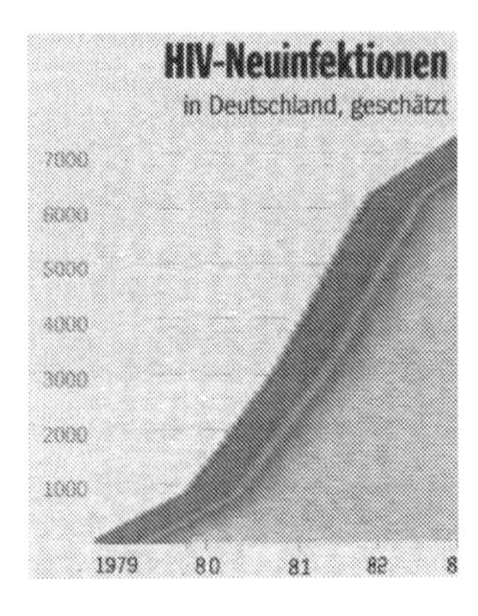

Figure 12.2

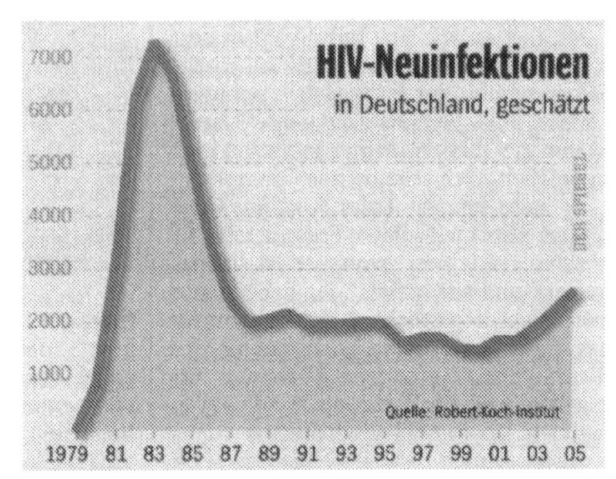

Figure 12.3

tween *models which predict* (for example the weather forecast), *models, which explain* (for example why do we see a rainbow), and *models, which describe* (for example the development of HIV).

The HIV example demonstrates how difficult it is to extrapolate from given data. Figure 12.2 shows data for HIV new infections in Germany from 1979 to 1983. This picture suggests a dramatic increase in numbers of HIV infections. Instead, as Figure 12.3 shows, the number of new infection decreased visibly after 1983, owing to improved precaution.

Models for a real problem can be more or less suitable. One should never talk about "right" or "wrong". For example, it does not make sense to call Newton's model of physics "incorrect" and Einstein's model "correct". Both models provide a reasonable description of nature under the given circumstances. There are "hard" models such as the quantitative models of physics which use model assumptions as the law of gravity. These models allow excellent predictions. Rather "soft" models such as models in economics or ecology, on the contrary, are often overvalued. In each case, models will always be of subjective character, owing to the normatively chosen assumptions. This also includes the danger of misuse and misinterpretation. It is an important task for school to impart knowledge about this fact.

BASIC EXPERIENCES IN MATHS EDUCATION

Following Heinrich Winter (1995/2004), the well-known German mathematics educator, three "basic experiences" are necessary in order that mathematics lessons will convey general education principles on every school level. Those basic experiences are

(BE 1) to realise in a specific way, and to understand phenomena in the world around us, which we are and should be concerned with,

(BE 2) to learn about and to understand mathematical issues represented in language, symbols, pictures, and formulas as intellectual creations, as a deductively ordered world of its own kind, and

(BE 3) to acquire problem-solving (heuristic) skills by analysis of tasks which go beyond mathematics.

The first basic experience points out the fundamental contributions of mathematics towards acquiring important knowledge about our world. The second one aims at the "inner world of mathematics", mathematics shows that a rigorous science is possible. And finally the third one describes mathematics as a school of thought. For example, "problem solving" (Henry Pollak: "Here is the problem, solve it") is inherent to the last basic experience.

Mathematics proves to be an inexhaustible pool of mathematical models, which allow us to understand better the world around us. However, for concrete examples, both of the two other basic experiences play central roles, too, especially when the important demand for interconnectedness according to the spiral principle is taken seriously. It is clear that our theme, modelling, relates to BE 1. In any case, applications and modelling are an indispensable part of the basic experiences of Winter and thus contribute to conveying a balanced image of mathematics in school at every level.

MODELLING: CHANCES AND OBSTACLES

Unfortunately, as a rule, reality-oriented teaching on applications outside mathematics is covered only to a limited extent in every day teaching although there is a long-standing agreement on the importance of creating relations between realistic situations and mathematics teaching.

I will especially point out three important factors which can promote the development of modelling competence of pupils, but can also obstruct them drastically. These are

- the problem field "central exams,"
- the use of computers, and
- the professional development and the motivation of the teachers.

The Problem Field "Central Exams"

Central examinations have a crucial influence on the content of teaching. Therefore, central examinations can also influence positively and foster the important relations between mathematics and "the rest of the world".

But often the so called application problems in central exams are only mathematical problems 'in disguise' and not genuine real life problems. For the students 'uncovering' these problems 'in disguise' is reduced to finding out the algorithms that have been hidden by the teacher, and immediately 'real' mathematics takes over. Those exams problems are more or less so called 'age-of-the-captain' problems. This kind of problems was first discussed from Stella Baruk (1989), such as the following:

> There are 26 sheep and 10 goats on a ship. How old is the captain?

Joe, a primary school pupil, picks both numbers and argues: $26 - 10 = 16$, that's the age of my brother, but he is too young to be a captain. $26 \cdot 10 = 260$, my grandpa is only 85, so that cannot be, too. 26 : 10 is not possible to calculate, so there is only one possible correct answer for the wanted age: $26 + 10 = 36$.

We laugh, but central exams often have the same impact: centralized assessment for final examinations or even tests at the end of each school year often reduce teaching to mindless drill and practice of procedures and calculation techniques. The following example is taken from the central final examination (Year 13) of the German state of Baden-Wuerttemberg, posed in 1998 in the topic Analytic Geometry: The problem is set in the context of a playground with a wooden pyramid that stands perpendicularly on a square base and is accessible inside.

The next text shows part c of the problem (my translation):

> Inside the pyramid a board is fixed parallel to the floor with a circular opening with a diameter of $d = 2.4$ in its middle. For tidying up, a big foam ball with radius $r = 1.5$ needs to be pushed through the opening towards the upper part of the pyramid. At which height is the board fixed if it is supposed to be as high up as possible with the ball lying loosely in the opening?

The missing measurement units show immediately that the problem poser does not take reality too seriously. Nowadays, it seems perhaps necessary to point out the importance of tidying up in school. Indeed, one puts more and more educational goals on the shoulders of the teachers. Back to the Abitur task: We assume that the measures are given in meter and make a drawing of the situation (Fig. 12.4).

The board is fixed at a height of 5.6 meters, and the ball possesses a volume of 9.4 cubic meters. The internet states the specific weight of foam:

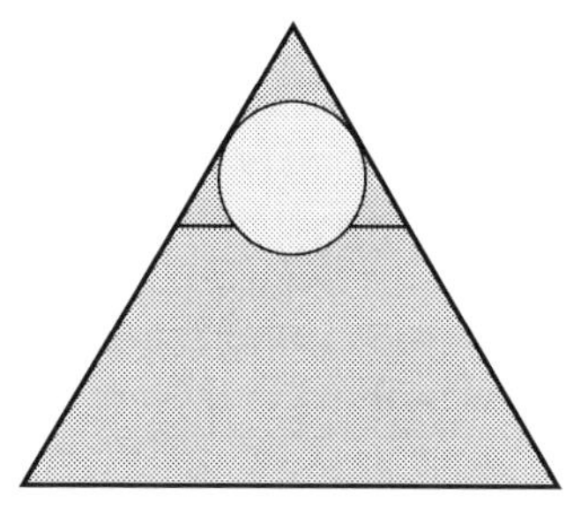

Figure 12.4

the ball weighs approximately 380 kg! How should this ball ever be pushed upwards? How should it ever be pulled out again? Maybe the problem poser was thinking about a giant screw pull like in Figure 12.5, to add the corresponding geometric helix curve to the problem?

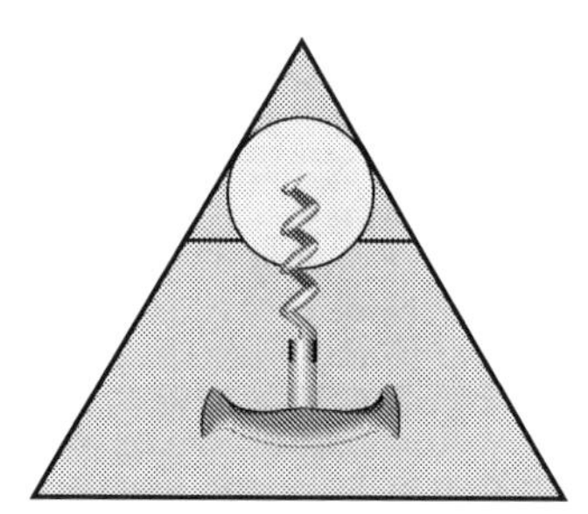

Figure 12.5

Anyway, the problem is a typical "age-of-the-captain" problem and such problems, given in central exams, influence teachers in their belief that modelling and applications are meaningless for mathematics teaching.

The Use of Computers

Today's available computer technology can contribute in a special way to aid in the learning process, and is equally helpful and important for all three basic experiences (see Henn, 1998). Firstly, the computer is a powerful tool to aid in modelling and simulation. Secondly, the computer can positively influence the generation of adequate basic concepts ("Grundvorstellungen") of mathematical ideas—especially through dynamical visualisations. Lastly, the computer furthers heuristic-experimental work in problem solving.

But the computer does what you want—sensible or foolish. The following problem, in my opinion, belongs to the last category. I found it in an American journal for didactics of mathematics (Martinez-Cruz & Ratcliff, 1998). The authors investigate the men's world record times in 100m freestyle swimming from 1912 to 1994. Without any mathematics, just using common sense, one would expect qualitatively something like the curve in Figure 12.6.

This qualitative curve has nothing to do with the modelling assumptions of logistic growth. For the intermediate time there are no reasonable model assumptions pointing at a special curve.

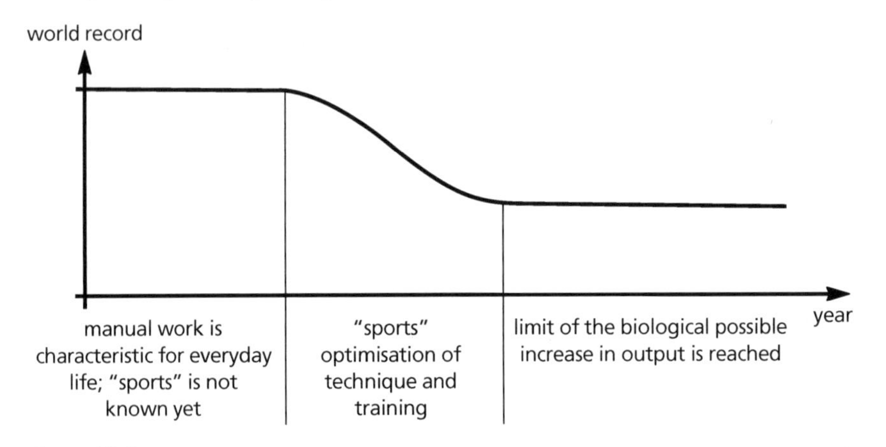

Figure 12.6

The authors use the following table (Table 12.1) of the world record times and fit various curves through the given record data by applying the regression commands available on their calculator.

TABLE 12.1

Year	Time (sec)
1912	61.6
1924	57.4
1957	54.6
1968	52.2
1972	51.22
1976	49.99
1988	48.42
1994	48.21

In detail, they fit a linear function $y = a \cdot x + b$, an exponential function $y = a \cdot b^x$, a power function $y = a \cdot x^b$, and a logistic function

$$y = \frac{c}{1 - e^{b+ax}}.$$

Figure 12.7 shows that the choice of the curve is irrelevant for the interval in question. However, extrapolation on both sides in Figure 12.8 shows that all models do not represent the real situation.

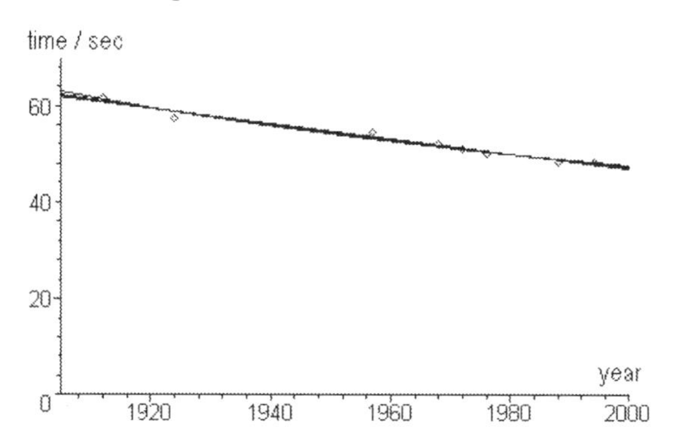

Figure 12.7

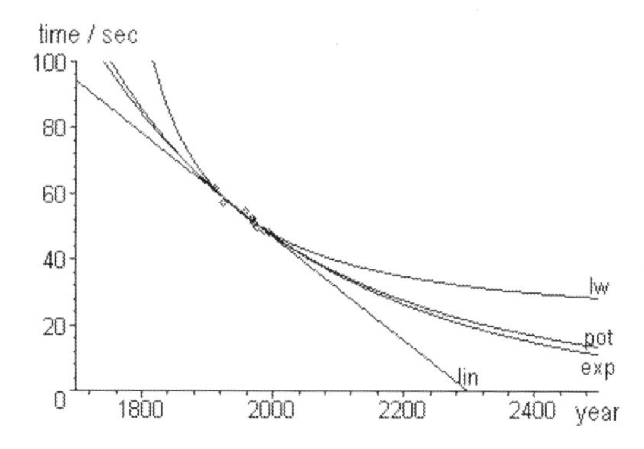

Figure 12.8

The authors favour the logistic model, because its predictions are less meaningless for the future compared to the other predictions! Naturally, the logistic curve—as often in occurrences from nature—is similar to the qualitative curve form one would expect, but it provides neither any explanation nor any meaningful prediction outside the measured data. No one of the four models provides deeper insight. The correlation coefficients of all four models do not differ significantly and are all nearly 1!

By the way, the authors do not consider one of the most interesting points of the table: it is the increase in measurement accuracy from 1968 to 1972. In 1972, the Olympic Games took place in Munich. At first, times were taken with three digits after the decimal point, an accuracy of 1/1000 second. This can be reconstructed from the results of the 400 m medley swimming contest (Table 12.2):

TABLE 12.2

400 m medley swimming (30.8.1972)	1. Gunnar Larsson (Sveden)	4:31.98 OR
	2. Tim McKee (USA)	4:31.98
	3. Andras Hargitay (Hungary)	4:32.70

At first, times were recorded as 4:31.981 for Larsson and 4:31.983for McKee and therefore, Larsson was awarded the gold medal. Then, obviously, somebody started thinking: It takes about 50 seconds to swim 100 m, that means a distance of about 2 mm in 1/1000 second. Nobody would believe that a 50 m long swimming pool could be constructed so accurately that each swimming lane had an accuracy of less than 2 mm. A little bit more mortar already leads to a larger difference. Therefore, the measurement accuracy was reduced to two digits after the decimal point. But, incomprehensibly no two gold medals were awarded!

The Professional Development and Motivation of Teachers

It is an important task to educate teachers to include applications and modelling in their teaching practice. This implies to "see the world with mathematical eyes" and to find occasions, again and again, to introduce some situations from reality in the mathematics classroom. A simple way to do this is to use newspaper clippings. However, caution must be applied not to overshoot the mark.

The following two newspaper clippings from different newspapers (from Herget & Scholz, 1998, my translation into English) denounce one of the often meaningless regulations which are issued from the German Federal Government.

Clipping 1:

Perfect official language

The perfection of German makers of regulations was documented by Secretary of the Interior Georg Tandler in the parliament in Munich by reading out the draft for an ordinance on holding calves in Germany. It is written there, however that might mean in plaintext: "If calves are held in herds each calf has depending on its withers height in centimetres to have a freely usable space in square meters according to the following formula: Minimal space (square cm) equals 0.4 times to the power 2 plus 70 times plus 2,720."

Clipping 2:

Authority Mathematics

"If calves are held in herds each calf has depending on its withers height in centimetres to have a freely usable space in square meters according to the following formula: (mathematical exponential way of writing) minimal space cm (to the power of) 2 equals 0.40 × (to the power of) 2 plus 70 × plus 2720." (Taken from the new draft of the German federal states for an ordinance on holding calves.) The CDU-member of the parliament of the German state of Hessen Dieter Weirich (Hanau) suggested to translate this draft to farmers and to give official help for computation. One needs to ask, commented Weirich, if farmers facing such "droppings from public offices" will ever get around to muck out their stables.

The two texts have mainly the same content, but there is one mathematically interesting difference in the way how each of the two authors describes the functional term:

1st text: "...following formula: Minimal space (square cm) equals 0.4 times to the power 2 plus 70 times plus 2,720."

2nd text: "...following formula: (mathematical exponential way of writing) minimal space cm (to the power of) 2 equals 0.40 × (to the power of) 2 plus 70 × plus 2,720."

Only the second journalist understood the mathematical symbol x correctly as the variable for the height of the calf, the first one reads it as the symbol for multiplication and gives a totally meaningless text. But even from the second text it is not easy to develop the correct formula. I gave the task to grade 9 students who developed about 6 or 7 formulas and wrote the down on the blackboard. Finally, we agreed on the following one.

$$f(x) = 0.40 \times x^2 + 70 \times x + 2{,}720.$$

After the graph has been drawn (Fig. 12.9) one can reflect on sense and nonsense of this regulation. One of my students immediately argued: It is much simpler to use a straight line!

The bold line in Figure 12.10 would give the same result but would lead to a simpler regulation.

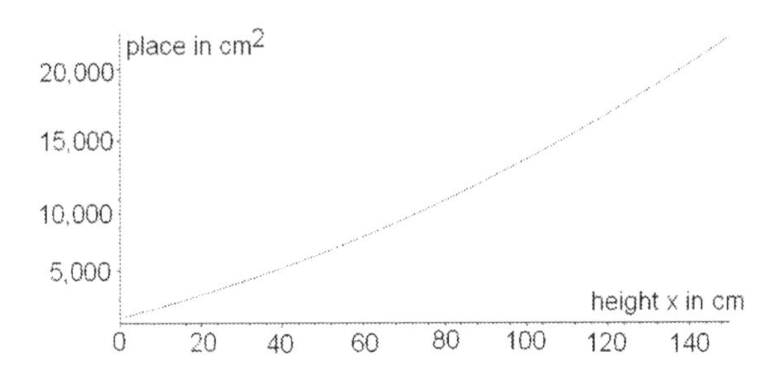

Figure 12.9

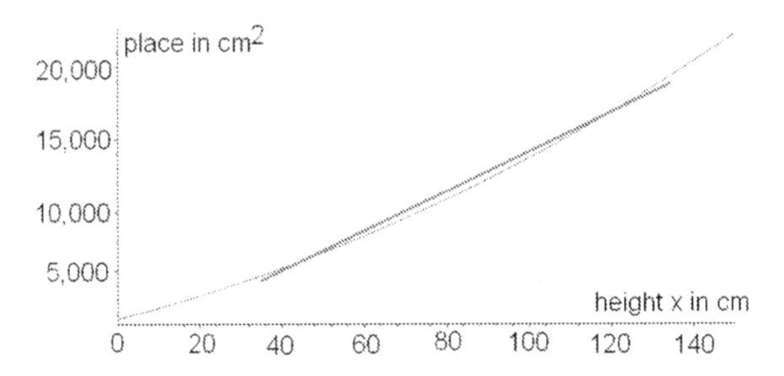

Figure 12.10

So far, so good! But, even this nice problem can be put into bad teaching practice out of sheer enthusiasm about applications and modelling. This is illustrated by the following two examples:

The first example is this schoolbook problem on the calf regulation in Figure 12.11 (from Sigma, 1984).

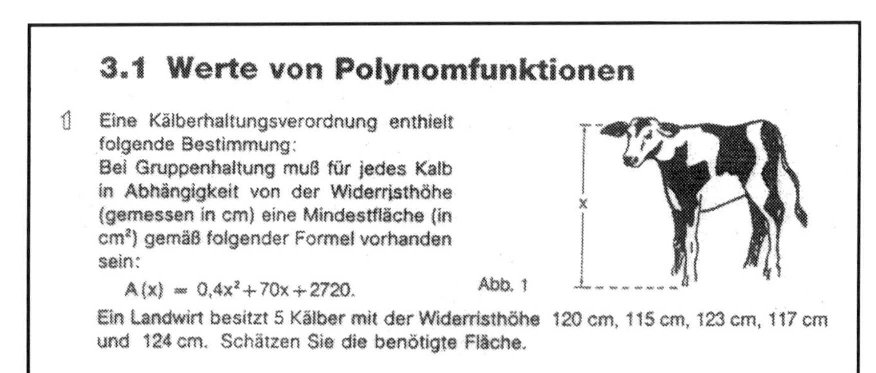

3.1 Werte von Polynomfunktionen

1 Eine Kälberhaltungsverordnung enthielt folgende Bestimmung:
Bei Gruppenhaltung muß für jedes Kalb in Abhängigkeit von der Widerristhöhe (gemessen in cm) eine Mindestfläche (in cm²) gemäß folgender Formel vorhanden sein:

$A(x) = 0{,}4x^2 + 70x + 2720.$ Abb. 1

Ein Landwirt besitzt 5 Kälber mit der Widerristhöhe 120 cm, 115 cm, 123 cm, 117 cm und 124 cm. Schätzen Sie die benötigte Fläche.

Figure 12.11

The topic is "Werte von Polynomfunktionen" (values of polynomial functions). At first, I will translate the task 1 into English:

An ordinance on holding calves contained the following provision: "If calves are held in herds each calf has depending on its withers height (measured in centimetres) to have a freely usable space (measured in square meters) according to the following formula:

$$A(x) = 0.4x^2 + 70x + 2720.$$

A farmer has 5 calves with withers heights 120 cm, 115 cm, 123 cm, 117 cm and 124 cm. Estimate the necessary area.

Without reflection, the regulation is cited, the term "Widerristhöhe," that means the withers height of the calves, is explained using a drawing with the variable x, instead having children search for explanations for themselves. Then, without any comment, the formula $A(x) = \ldots$ is given. The task is now to substitute five values in the formula and to add. This is not the way to develop a reality-oriented problem, but a typical age-of-the-captain problem.

The second example is the following part of the mathematical diary of a girl (my translation into English). A young teacher, more correct, a teacher trainee, had covered the calves problem adequately in the classroom and now the girl reports on this:

I owe the second page to Mrs. Koch, a teacher trainee.

The problem was set in the context of calculating the necessary space in a stable for a calf of size x. Maybe this was meant to broaden the pupils' horizon for the unlimited possibilities to use functions. In this case the function increased exponentially, which would mean that the farmer needed to apply a straightedge regularly to find out about the growth of each of the calves and then to assign them a new, bigger place in the stable. I would argue in favour of a minimal value that would make any calculation superfluous.

However, according to Mrs. Koch, a linear function would turn out to be an indispensable help for the farmer, because he could read off the necessary space comfortably from a graph.

I do not agree to this. How would his life be made easier, if his stable needed to look like this?

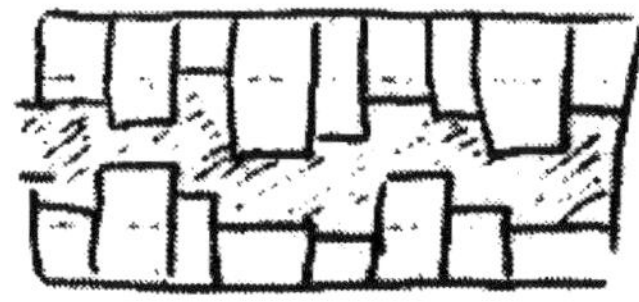

Of course, the graph does not "increase exponentially" as the girl mentioned, but that is not the point. The point is that the teacher, out of pure enthusiasm, gave the impression that a regulation with a linear formula would be the only reasonable solution. The girl proved to have more common sense than the teacher.

MODELLING IN SCHOOL

Mathematical modelling is the mutual fertilization of mathematics and the rest of the world (Pollak, 1979). Through modelling, students are enabled to build a bridge between mathematics as a tool to understand better the world around them, and mathematics as abstract structure. For this, suitable teaching situations are indispensable. Lyn English (2003) demands "rich learning experiences", i.e. authentic situations, chances for own exploration, multiple possibilities for interpretations, and social competence to take up the responsibility for one's own model up to communicating it to other students. Teachers are often reluctant to include mathematical modelling in their teaching. Katja Maaß (2006) points out that the complete modelling process is time-consuming and difficult. She also shows, however, that modelling activities can be started successfully in a normal teaching situation. Students should not be spared the difficulties and effort related to applications and modelling. I will demonstrate this with some examples.

Who Needs Numbers, Has the Choice: Ideal, Real, and Computer Numbers

From my point of view, the number line is the most important model in school. Meaningful use of numbers is an indispensable skill for educated citizens. In life, in and after school, numbers happen to occur in three forms: There are the ‚ideal' numbers of mathematics, for which obviously

$$2 = 2.0 = 2.00$$

applies, and the real numbers from daily life, which often turn out to be intervals, for which 2 will not be the same as 2.0. Mostly, for example when measuring, intervals, not exact numbers are an adequate model for the situation. Intervals, however, will lead to error propagation in further calculations. If this is forgotten, results can become arbitrary very quickly. On top, nowadays we also have the computer numbers, which lead a life of themselves. While the performance of processors has increased rapidly, the error analysis of the implemented floating point arithmetic has been neglected immensely.

A quick estimation illustrates the rounding problems of numerical calculations: The mathematical model of the reflection law states for each angle of incidence α always the exact value $\beta = \alpha$ of the angle of reflection (Fig. 12.12).

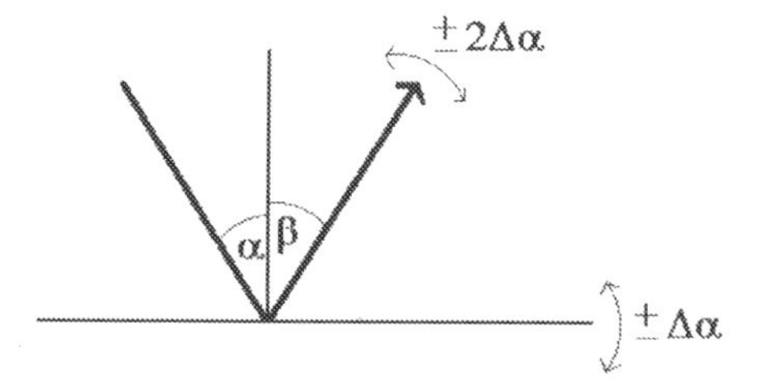

Figure 12.12

But, if the position of the reflector is only known with a deviation of $\pm\Delta\alpha$, then the angle of reflection β is given by the interval $[\alpha - 2\Delta\alpha; \alpha + 2\Delta\alpha]$, the error has doubled. After n reflections, the error has increased to $\pm 2n \times \Delta\alpha$ Even with a very small initial error of 1/1000 degree the error has increased to a value of more than 360° after 18 reflections—no predictions can be made any more.

Examples from Practice for Practice

The following few examples show that "seeing the world with mathematical eyes" will provide many occasions to motivate modelling activities in the classroom.

Maintenance House

Figure 12.13 shows a maintanence house along a railway line. It is not clear, how the distance indication can be given exactly in meters. The picture provides a good starting point to discuss the use of numbers which should be suitable to describe a real situation.

Figure 12.13

USA Eagle Stamp

In a publication for the European market the US postal administration listed the size of the Eagle Stamp (Fig. 12.14) released on April, 29th, 1985, to be 48.768 × 43.434 millimeter.

Figure 12.14

One little fault is that the measurement unit should be square millimeters. Again meaningless is the exactness of the measures. Already a change of humidity will change the size of the stamp more! What is the reason? In the United States the size of the stamp was listed to be 1.92 × 1.71 square inches. The person responsible for converting the size to metric measurements for the European market simply multiplied with the conversion factor of 2.54 and forgot to round meaningfully. A further example for lacking numeracy!

Mount Everest and Other Mountains

While touring Britain we travelled near Ben Nevis, the highest mountain in Scotland. The guide said she had no clue how one could determine the height of this mountain. Again and again it is amazing that even educated persons have no idea how to do this, even if intercept theorems and trigonometry are taught in school. Memorized formulas and the application of mathematical theorems to concrete application problems are two different things. Activities doing concrete measurements outside with self-built measuring instruments should be an integral part of teaching in the middle grades.

There is another mathematically interesting story about the measurement of the height of the highest mountain, the Mont Everest (Fig. 12.15).

Figure 12.15

Around the middle of the 19th century British surveyors took measurements on the height of the mountain from different places (Poindexter, 1999). Taking the arithmetic mean from the different measures in 1852, by chance they got a result of exactly 29,000 British ft. They did not want to publish this result as they worried this would be taken as an imprecise measurement with an accuracy of only ±1,000 ft. Therefore, they published 29,002 ft as the height of the Mount Everest.

Small Sugar Bags

In restaurants one gets sugar in various packages (Fig. 16). Of special interest are the bags provided by Burger King (Fig. 17), which have an imprint stating that "45 % less paper than usual packing" is used.

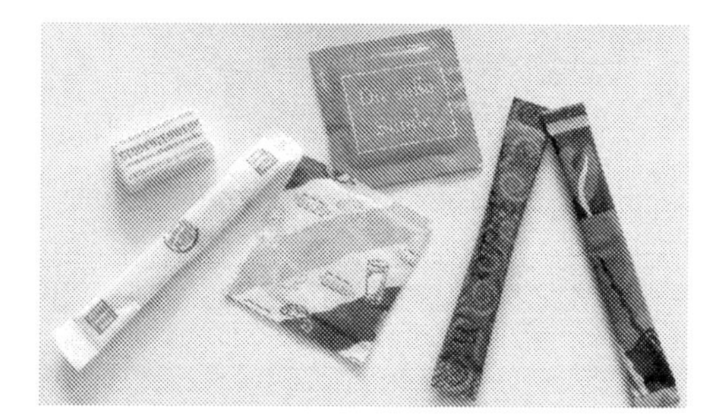

Figure 12.16

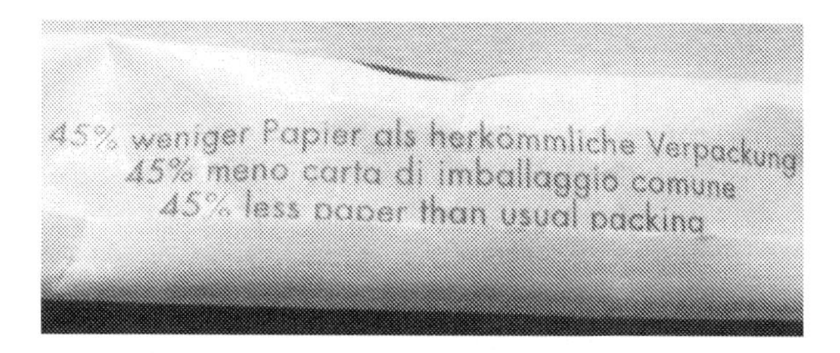

Figure 12.17

Secondary school students can investigate and judge on this statement using various methods.

CONCLUSION

Teaching affects the image that students will take with them into their future life as responsible citizens and future decision makers. This image should contain both the beauty *and* the functionality of mathematics. But applications of mathematics in other fields should not be studied for its own purpose alone. Reflecting on what relates mathematics with the rest of the world is indispensable, ethical issues of mathematical actions have to be highlighted, and students have to be sensitised for it.

REFERENCES

Baruk, S. (1985). *L'âge du capitaine. De l'erreur en mathematiques.* Paris: Editions du Seuil.

Blum, W., Henn, H.-W., Galbraith, P., & Niss, N. (2007). *Modelling and Applications in Mathematics Education.* New York: Springer 2007.

Englisch, L. (2003). Mathematical Modelling with Young Learners. In Lamon, S. J. et al. (Eds), *Mathematical Modelling: A Way of Life,* (pp. 3–17). Chichester: Ellis Horwood.

Henn, H.-W. (1998). The Impact Of Computer Algebra Systems On Modelling Activities. In Galbraith, P, W. Blum, G. Booker, & I. Huntley (Eds.), *Teaching and Assessing in a Technology rich World,* (pp. 115–123). Chichester: Ellis Horwood.

Herget, W. & D. Scholz (1998). *Die etwas andere Aufgabe—aus der Zeitung.* Seelze: Kallmeyer.

Maaß, K. (2006). Modelling in classrooms: What do we want the students to learn? In Haines, Ch. et. al. (Eds.), Mathematical Modelling (ICTMA 12): Engineering and Economics. Chichester: Ellis Horwood.

Martinez-Cruz, A.M. & Ratcliff, M.I. (1998). Beyond modeling world records with a graphing calculator: Assessing the appropriateness of models. *Mathematics and Computer Education 32*/2, 143–153.

Neubrand, M. (2003). "Mathematical Literacy"/"Mathematische Grundbildung": Der Weg in die Leistungstests, die mathematikdidaktische Bedeutung, die Rolle als Interpretationshintergrund für den PISA-Test. Zeitschrift für Erziehungswissenschaften 6 (3), 2003, S. 338–356

Niss, M. (1994) Mathematics in Society. In Biehler, R., Scholz, R. W., Straesser, R., Winkelmann, B. (Eds.), *The Didactics of Mathematics as a Scientific Discipline.* Dordrecht: Kluwer, p. 367–378.

Pollak, H. (1979). The interaction between mathematics and other school subjects. In UNESCO (Ed.): *New Trends in mathematics teaching, Vol IV.* Paris, pp. 232–248.

Poindexter, J. (1999). *Zwischen Himmel und Erde.* Köln: Könemann.

Pratchett, T. (2005). *Das Erbe des Zauberers.* München: Piper-Verlag.

Sigma Mathematik 11. Klasse (1984). Stuttgart: Klettverlag

Winter, H. (1995/2004). Mathematikunterricht und Allgemeinbildung. *Mitteilungen der Gesellschaft für Didaktik der Mathematik,* Nr. 61, Dez. 1995, S. 37–46. Revised version in H.-W. Henn, & K. Maaß (Eds.), *ISTRON Materialien für einen realitätsbezogenen Mathematikunterricht, Band 8.* (pp. 6–15). Hildesheim: Franzbecker.

CHAPTER 13

COMPLEX SYSTEMS IN THE ELEMENTARY AND MIDDLE SCHOOL MATHEMATICS CURRICULUM

A Focus On Modeling

Lyn D. English
Queensland University of Technology, Australia

Abstract: An appreciation and understanding of the world as comprising interrelated complex systems is critical for all citizens in making effective decisions about their lives, their families, their communities, and the environment. This paper argues for the need to introduce elementary and middle school children to the rudiments of "basic complex" systems within the mathematics curriculum. A powerful avenue for doing so is through mathematical modeling. Following an overview of the key features of complex systems and mathematical modeling, a sample problem is presented to illustrate one way in which children can engage with basic complex systems. Some ways in which 4th and 5th grade children solved the problem are presented.

Studies of complex systems have proliferated the literature in the past decade (e.g., Bar-Yam, 2004; Hmelo-Silver, Marathe, & Liu, 2007; Jacobson,

The Montana Mathematics Enthusiast, pages 177–196

2001; Jacobson & Wilensky, 2006; Lesh, 2006). Calls for school curricula to increase students' exposure to complex systems are clearly warranted, given that we live in a world that is increasingly governed by such systems: the World Wide Web, political parties, financial corporations, school and university systems, family structures, and sporting teams are just a few examples. An appreciation and understanding of the world as comprising interlocked complex systems is critical for all citizens in making effective decisions about their lives as both individuals and as community members (Bar-Yam, 2004; Jacobson & Wilensky, 2006; Lesh, 2006).

Research addressing students' understanding of complex systems has been confined largely to the secondary and undergraduate years, with the domain of science being a popular field of investigation (e.g., see Jacobson & Wilensky's review, 2006). Studies of complex systems have had virtually no impact on mathematics education (English, 2007a). This is not because mathematics educators have been ignorant of such developments; rather, current learning science theories do not provide sufficient answers to how we can promote students' understanding of complex, real-world systems that permeate their life. As Lesh (2006) noted in the complex systems strand of the *Journal of the Learning Sciences* (15, 1):

> I do not view current learning science theories as being sufficient to provide answers to most questions about the nature of the conceptual systems that students would need to develop to understand complex systems. In contrast, I believe that the most exciting point about learning science investigations of complex systems is precisely that such research is likely to require a variety of significant paradigm shifts beyond current ways of thinking. Furthermore, I believe that these paradigm shifts should have implications for learning and problem solving related to a wide range of constructs and situations where relationships to systemic understandings are far less obvious than in the case of complex systems (p. 46).

In this paper, I add further support to Lesh's (2006) argument that current complex systems research is not sufficient for addressing the nature of the conceptual systems that students need in dealing with "everyday" complexity in their world. What this research is providing, however, are important theoretical shifts in how we can address complex problem solving and learning in the mathematics curriculum. I present here an alternative perspective on developing students' facility with complex systems. My focus is not on the "deeply complex" systems that involve many layers of complexity where the "agents" within the system are often living organisms or ecosystems (Lesh, 2006, p. 47). Rather, I consider "basic complex" systems that are inherent in authentic mathematical modeling problems designed for elementary and middle school students. In order to appreciate the nature of these systems, I first review briefly some of the key features of deeply complex systems. I then identify the main components of basic complex systems, which I have been introducing to elementary and middle school students through mathematical modeling (e.g., English, 2007b). Following an overview of modeling, I provide an example of a problem involving a basic complex system and present some examples of how fourth- and fifth-grade children worked the problem.

"DEEPLY COMPLEX" SYSTEMS

> Understanding complex systems is fundamental to understanding science. The complexity of such systems makes them very difficult to understand because they are composed of multiple interrelated levels that interact in dynamic ways. (Hmelo-Silver et al., 2007, p. 307)

As previously noted, studies that have explored students' understanding of complex systems within the biological sciences have been popular (e.g., Hmelo-Silver et al., 2007; Hmelo-Silver, Holton, & Kolodner, 2000). This is not surprising, given the many instances of deep complexity that abound in this domain. For example, processes that take place within organisms, the interactions between organisms and their environment, and the interactions among various organisms within and across species are addressed in terms of complex systems (Hmelo-Silver et al., 2007). Although such studies have some import for mathematics education, as I indicate here, they nevertheless address a more advanced level of complexity than is warranted in the elementary and middle school mathematics curriculum.

Defining complex systems per se has been a challenge in itself. Over a decade ago Horgan (1995) published an article titled, *From Complexity to Perplexity*, in which he highlighted the struggle of researchers to find a unified definition. One basic definition is that complex systems comprise sets of interconnected elements (parts) whose collective behaviour arises in an often counterintuitive and surprising way from the properties of the elements and their interconnections (Jakobsson & Working Group 1 Collaborators, http://necsi.org/events/cxedk16/cxedk16_1.html). For example, if we consider a complex system such as proteins, we have their elements (e.g., amino acids), their interactions (e.g., bonds), their formation (e.g., protein folding), and their activity (e.g., enzymatic activity; Bar-Yam, 1997). The properties, however, only become meaningful within the context of a system functioning as-a-whole; in other words, the whole is more than the sum of its parts (Lesh, 2006). At the same time, we cannot just focus on the whole; we need to move back and forth between seeing the parts and the whole. That way, we can see which aspects of the parts are relevant to describing the whole. This relationship between the parts and the whole is referred to as *emergence* (Bar-Yam, 2004).

Another key feature of the relationships between parts and wholes is *interdependence*, namely, a consideration of how the parts of a system affect one another (Bar-Yam, 2004). Here the focus is on the strength of the dependencies between the parts, and how manipulating or removing a particular part will affect the rest of the system. Sometimes the effect is small and at other times, large, and there might be many effects on the system or only a few. These different kinds of interdependence are a major consideration in our efforts to solve a variety of problems, ranging from family issues through to major global concerns. It is thus important that students be given opportunities to investigate these different interdependencies in their efforts towards understanding relatedness and relationships. The challenge for the mathematics education curriculum is how to integrate experiences involving complex systems that introduce students to some of these fundamental features. One approach I suggest here is through experiences with basic complex systems (or "simply complex" using Lesh's term, 2006, p. 47), which deal with meaningful real-world situations that engage students in generating significant mathematical constructs and processes.

"BASIC COMPLEX" SYSTEMS

Most of the complex systems that students encounter in their daily lives are really only complex at the highest level, with their component parts following rather simple functional rules (Lesh, 2006). This is in contrast to the deeply complex systems found in the biological sciences, where additional layers of complexity tend to emerge because the living organisms within the system are not generally governed by simple rules. Despite these differences, the two systems share core features. Those features that pertain to the basic complex systems I address here include:

- Sets of interconnected parts
- Interactions among the parts
- A system functioning as-a-whole
- Property of emergence: relationship between the parts and the whole
- Property of interdependence: how the parts of the system affect one another.

To introduce children to basic complex systems, I have been implementing mathematical modeling problems in elementary and middle schools (e.g., Doerr & English, 2003; English, 2007b; English, 2006; English & Watters, 2005). These problems, as I indicate later, normally display the above features and provide students with experiences in dealing with "everyday" complex systems. Prior to addressing an example of such experiences, I provide some background information on mathematical modeling for the elementary and middle school.

MATHEMATICAL MODELING FOR THE ELEMENTARY AND MIDDLE SCHOOL

Mathematical modeling is foundational to modern scientific research, such as biotechnology, aeronautical engineering, and informatics. Because mathematical models entail a focus on the structural characteristics of phenomena (e.g. patterns, interactions, and relationships among elements) rather than surface features (e.g. physical or artistic attributes), they are powerful tools in predicting the behaviour of complex systems (Harel & Lesh, 2003).

The terms, *models* and *modelling*, have been used variously in the literature, including with reference to solving word problems, conducting mathematical simulations, creating representations of problem situations (including constructing explanations of natural phenomena), and creating internal, psychological representations while solving a particular problem (e.g., Doerr & Tripp, 1999; English & Halford, 1995; Gravemeijer, 1999; Greer, 1997; Lesh & Doerr, 2003; Romberg et al., 2005; Van den Heuvel-Panhuizen, 2003). The perspective that I have adopted in my research is that of models as "systems of elements, operations, relationships, and rules that can be used to describe, explain, or predict the behavior of some other familiar system" (Doerr & English, 2003, p.112).

Mathematical modeling takes children beyond the usual form of problem solving they meet in the elementary school. Typical classroom problems present the key mathematical ideas "up front" and children select an appropriate solution strategy to produce a single, usually brief, response. In contrast, modeling problems have the important mathematical constructs and relationships embedded within the problem context and children elicit these as they work the problem. The problems necessitate the use of important, yet underrepresented, mathematical processes such as constructing, describing, explaining, predicting, and representing, together with organizing, coordinating, quantifying, and transforming data—all of which are important for dealing with complex systems in our world (e.g., Gainsburg, 2006; Lesh & Doerr, 2003). The problems are also multifaceted and multidisciplinary: students' final products encompass a variety of representational formats including written text, graphs, tables, diagrams, spreadsheets, and oral reports; the problems also cut across several disciplines including science, history, environmental studies, literature, and history.

Importantly, these modelling problems encourage the development of generalizable models. In my research with elementary and middle school children, I have implemented sequences of modeling problems that encourage the creation of models that are applicable to a range of related situations (e.g., Doerr & English, 2003; Doerr & English, 2006; English & Watters, 2005). Children are initially presented with a problem that confronts them with the need to develop a model to describe, explain, or predict the behavior of a given system (a model-eliciting problem). Given that re-using and generalizing models are central activities in a modelling approach to learning mathematics and science, the children then work related problems that enable them to extend, explore, and refine those constructs developed in the initial problem (model-exploration and model-application problems; Lesh, Cramer, Post, & Zawojewski, 2003). Because the children's final products embody the factors, relationships, and operations that they considered important, powerful insights can be gained into the children's mathematical thinking as they work the problem sequence.

Finally, unlike typical school problems, modeling activities are social experiences (Zawojewski, Lesh, & English, 2003), where children engage in small-group collaborative work and are motivated to challenge one another's thinking and to explain and justify their ideas and actions. Numerous questions, issues, conflicts, revisions, and resolutions arise as children develop, assess, and prepare to communicate their solutions. The ability to work collaboratively on comprehensive problems in which planning, monitoring, and communicating results are essential to success is not normally developed in typical classroom problem solving; dealing effectively with complex systems calls for such skills.

AN EXAMPLE OF A MODEL-ELICITING PROBLEM INVOLVING A BASIC COMPLEX SYSTEM

"Christmas Dinner at the Thompson's" (Appendix A) is a model-eliciting problem that I presented to 55 fourth- and fifth-grade students (9- and 10-year-olds) who attended a two-day "Mathematical Modeling Challenge" for schools in south-eastern Queensland, Australia. The children worked the problems in groups of three to four during the second day of the Challenge activities (3 sessions totalling approx. 3 hours were devoted to the Dinner problem). After the children had solved the problem, they presented group reports on their models and their creations to the class. The reporting session was an important aspect of the problem activity as it allowed the children to compare and contrast ideas, to provide constructive feedback to their peers, and to build on ideas to create new understandings (cf. notion of knowledge building, Bereiter & Scardamalia, 2006).

As the children worked the problem, the research assistants and I moved around the classroom observing the children's interactions and, where appropriate, asked them to explain or justify their ideas. No direct teaching was presented; if the children sought assistance we would advise them to seek answers within their group and, if necessary, would pose one or two questions to refocus their thinking. We videotaped and audiotaped both the children's conversations and actions, and their final class presentations.

The Christmas Dinner problem comprises several components: background information (in text and table format) that sets the problem context and alerts the students to some factors they need to consider in working a problem of this nature (Part A); "readiness questions" (not included in Appendix A) designed to ensure the students have understood the information presented; additional data and information for consideration in working the problem (part B); and the problem "challenge" itself (Part C).

As can be seen, the problem is multifaceted and comprises (a) sets of interconnected parts (e.g., sets of various ingredients; food items to be oven baked; components of cleaning and decorating; responsibilities and tasks to be shared among the children) and (b) interactions among these parts (e.g., how and when the food items will be placed in the oven and what will be placed with what [and who will oversee this]; how cleaning and food preparation will be integrated; how the available bench space will govern food preparation). The overall system (schedule) must function as a whole for the dinner to be a success. For example, the three children need to coordinate all of their actions efficiently, taking into consideration the various tasks that have to be done and when they have to be done, the availability of needed utensils and bench space, and the timing of each task. As the children develop and implement their system, they need to assess how the various parts are working towards creating the whole (the dinner) [emergence property]. They also need to consider how the various parts affect one another (interdependence). For example, in preparing their parents' anniversary dinner (Part A), the children overlooked how the availability of kitchen equipment affected the timing of the different tasks.

HOW CHILDREN DEALT WITH THE BASIC COMPLEX SYSTEM IN THE DINNER PROBLEM

In this section I present a few group reports that the children presented to their class peers on completion of the Dinner problem. Their reports indicated the children's awareness of the multifaceted nature of the problem, with the need to consider each of the different parts, their interactions and interdependence, and their functions in forming the overall system.

The "Marvellous Mathematicians" group explained their reasoning in creating their schedule as follows. The actual schedule they developed appears in Appendix B.

> This activity is all about organisation and planning. The challenge we are faced with is to organise a Christmas Dinner schedule for the Thompson family kids. They are doing this for their hard-working parents to make their limited holiday time relaxing and enjoyable. They have made a dinner before but they made a mistake in the planning and all, and all needed to use the oven at the same time. In our schedule we need to include, who uses what, when and for how long. We also need to include smaller chores and cleaning time.

When asked for further explanation of the development of their schedule, the children stated:

> We started by figuring out what time we wanted to serve each different dish. Then from that information we could figure out what time Dan, Eva, and Sophie needed to start cooking the different dishes so that they would be ready in time. Next we made a schedule showing who cooked what and when. We also made an extras chart showing who did the cleaning, decorations and other small tasks.

I followed up by asking the group to identify some of the aspects of the problem that presented a challenge to them. They explained:

> Well sometimes we kept getting pressed for time; they needed to do two things at the same time so we had to fix that... Um, like we just made them do all the activities and in the end it did work out.... sometimes [we] had to switch a few things round.... cause sometimes there would have been like half an hour when they weren't doing anything so we moved everything down a bit of time

Another group, the "Magnificent Mathematics Monkeys" stressed the importance of developing an effective schedule and explained how they followed a pattern in achieving this:

> Dan, Sophie and Eva have decided to make a delicious dinner for their hard-working parents. The most important part of the dinner is the schedule. Without it the night would be a wreck. Our task as a group was to create a schedule that if followed would have the night run smoothly. Our group started off with the timetable for the most obvious facts and then we proceeded to add information where it was needed. As we progressed with the project we started to get into a pattern. We would do two activities and then we would clean up in

the kitchen. Using this pattern we um, this pattern and other solutions we successfully finished the project.

When asked to explain their "pattern," the children responded:

> Um, well when we started working, instead of doing one task and then cleaning up afterwards we found that we saved more time doing two tasks and then cleaning up after that. So we were saving time when we did that.

A third group, the "Banana Splits" highlighted the elements they took into consideration in developing their schedule and also noted that their system would be applicable to other related problem situations. They explained their procedures as follows:

> How we came up with the schedule. To make the schedule we considered the time it takes to cook it, what utensils to cook each thing, but we also took into consideration whether they can cook anything else and whether each thing is warm or cold. With all these elements and a bit of trial and error we came up with our solution. This schedule is reusable for other events because of the list of elements above. Enjoy!

On asking the group to elaborate on the important components that made their system reusable, the children explained:

> How long it takes to cook it, what um, dishes were needed to cook each thing, like what utensils, and we also took into consideration whether they could cook with anything else at the same time, and yeah, and if they were served warm or cold.

On conclusion of the group reports, we asked the children to reflect on the factors that they considered important in developing their schedule. The following discussion took place:

> Andy: Time as in how long it takes to cook, how long it takes to prepare, how long everything's going to take and how much time you actually have...

We asked if time was "one of the first things you needed to consider?" Lachlan responded:

> Yes you have to consider how much time you have to cook everything cause if you don't get it in at the right time, its going to be like, say main course, entrée, dessert, something like that!

Other children commented that one also needs to know: (a) "When things are going to be served," (b) that "You have to use all three children,"

(c) that "You have to know what utensils and things you need and how much space you have or how much oven space you have," and (d) "You've got to make sure you have the ingredients necessary, not like go to make it and expect you have all the ingredients and go out and buy them."

In a follow-up discussion on other situations in which they could apply their model, the children commented that they could apply it to other occasions such as birthdays, Easter lunches, and anniversaries. In extending their thinking here, we asked for some different situations. James responded:

> Planning for a competition, like a touch [football] competition, it's not quite the same but you're still planning it, like what time that event is on, like that team and that team are versing [opposing] each other at that time...and how long the competition would go and then like quarter finals, semi-finals...

Another child suggested that their model could be applied to "bus and train timetables, with a schedule for when trains and buses are going to leave..."

DISCUSSION AND CONCLUDING POINTS

This paper has argued for the inclusion of experiences with complex systems in the elementary and middle school mathematics curriculum. Although there has been substantial research on students' understanding of complex systems, especially in the biological sciences, such research has had almost no impact on the mathematics curriculum. However, studies of complexity are providing us with new ways in which we can structure problem-solving experiences that expose children to everyday complex systems. I have proposed here the notion of "basic complex" systems as a way of introducing children to examples of complexity in their world. This notion differs from "deeply complex" systems that involve many layers of complexity and usually involve living organisms or ecosystems.

Basic complex systems have formed the basis of the mathematical modeling problems I have introduced to elementary and middle school students. These systems share many of the features of the deeply complex systems: they comprise sets of interconnected parts with interactions among the parts; the systems function as a whole; and they display the properties of emergence (relationships between the parts and the whole) and interdependence (how the parts of the system affect one another).

The modeling problems involving basic complex systems that I have designed and implemented in elementary and middle school classrooms extend children's problem-solving experiences beyond their usual curriculum. Rather than being presented with "neatly packaged" mathematics, as in standard word problems, children have to elicit the important mathematical ideas and processes themselves as they work the problems. That is, the problems require children to make sense of the situation so that they can mathematize it themselves in ways that are meaningful to them.

In the Christmas Dinner problem presented here, children dealt with some of the key features of basic complex systems as they developed a schedule for organizing a family function. The children's class presentations and written artefacts displayed their awareness of the comprehensive nature of the problem and the need to consider each of the different parts, their interactions and interdependence, and their functions in creating the system itself. The children could also identify other situations in which they could apply their models.

Mathematical modeling provides a rich avenue through which children can learn about basic complex systems. Because they regularly encounter such systems in their everyday lives, children need more exposure to mathematical problems that equip them with skills to work with complex systems. The inclusion of such problems in the regular mathematics curriculum is paramount. The level of complexity children experience in their world is increasing rapidly—we need to ensure they can deal effectively with this complexity.

REFERENCES

Bar-Yam, Y. (1997). *Dynamics of complex systems.* Reading, MA: Addison-Wesley.

Bar-Yam, Y. (2004). *Making things work: Solving complex problems in a complex world.* NECSI: Knowledge Press.

Bereiter, C., & Scardamalia, M. (2006). Education for the knowledge age: Design-centered models of teaching and instruction. In P. A. Alexander & P. H. Winne (Eds.), *Handbook of educational psychology* (2nd edition; pp. 695–714). Mahwah, NJ: Lawrence Erlbaum.

Davis, B., & Simmt, E. (2003). Understanding learning systems: Mathematics education and complexity science. *Journal for Research in Mathematics Education, 34*(2), 137–167.

Davis, B., & Sumara, D. (2006). *Complexity and education: Inquiries into learning, teaching, and research.* Mahwah, NJ: Lawrence Erlbaum Associates.

Doerr, H., & English, L. D. (2006). Middle-grade teachers' learning through students' engagement with modelling tasks. *Journal for Research in Mathematics Teacher Education, 9*(1), 5–32.

Doerr, H. M., & English, L. D. (2003). A modeling perspective on students' mathematical reasoning about data. *Journal for Research in Mathematics Education, 34*(2), 110–137.

Doerr, H. M., & Tripp, J. S. (1999). Understanding how students develop mathematical models. *Mathematical Thinking and Learning, 1*(3), 231–254.

English, L. D. (2007a). Cognitive psychology and mathematics education: Reflections on the past and the future. In B. Sriraman (Ed.), *Zoltan Paul Dienes and the dynamics of mathematical learning. The Montana Mathematics Enthusiast, Monograph No. 2* (pp. 119–126). Missoula: The University of Montana and the Montana Council of Teachers of Mathematics.

English, L. D. (2007b). Interdisciplinary modelling in the primary mathematics curriculum. In J. Watson & K. Beswick (Eds.), *Mathematics: Essential research, essential practice* (pp. 275–284). Hobart: Mathematics Education Research Group of Australasia.

English, L. D. (2006). Mathematical modeling in the primary school: Children's construction of a consumer guide. *Educational Studies in Mathematics, 62*(3), 303–329.

English, L. D., & Watters, J. J. (2005). Mathematical modeling in the early school years. *Mathematics Education Research Journal, 16* (3), 58–79.

English, L. D., & Halford, G. S. (1995). *Mathematics education: Models and processes.* Mahwah, New Jersey: Lawrence Erlbaum Associates.

English, L. D., Fox, J. L., & Watters, J. J. (2005). Problem posing and solving with mathematical modelling. *Teaching Children Mathematics, 12*(3), 156–163.

Gainsburg, J. (2006). The mathematical modeling of structural engineers. *Mathematical Thinking and Learning, 8*(1), 3–36.

Goldstone, R. L. (2006). The complex systems see-change in education. *The Journal of the Learning Sciences, 15* (1), 35–44.

Gravemeijer, K. (1999). How emergent models may foster the construction of formal mathematics. *Mathematical Thinking and Learning, 1,* 155–177.

Greer, B. (1997.) Modeling Reality in Mathematics Classroom: The Case of Word Problems. *Learning and Instruction 7,* 293–307.

Harel, G., & Lesh, R. (2003). Local conceptual development of proof schemes in a cooperative learning setting. In R. Lesh & H. M. Doerr (Eds.). (2003). *Beyond constructivism: Models and modeling perspectives on mathematic problem solving, learning and teaching* (pp. 359–382). Mahwah, NJ: Lawrence Erlbaum.

Hmelo-Silver, C. E., Marathe, S., & Liu, L. (2007). Fish swim, rocks sit, and lungs breathe: Expert-novice understanding of complex systems. *Journal of the Learning Sciences, 16*(3), 307–331.

Hmelo-Silver, C. E., & Azevedo, R. (2006). Understanding complex systems: Some core challenges. *The Journal of the Learning Sciences, 15*(1), 53–62.

Hmelo-Silver, C. E., Marathe, S., Liu, L. (2004, June). *Understanding complex systems: An expert-novice comparison.* Paper presented at annual meeting of the American Psychological Association, Honolulu, HI.

Hmelo-Silver, C. E., Holton, D. L., & Kolodner, J. L. (2000). Designing to learn about complex systems. *The Journal of the Learning Sciences, 9* (3), 247–298.

Horgan, J. (1995). From complexity to perplexity. *Scientific American, 272*(6), 104–110.

Jacobson, M. J. (2001). Problem solving, cognition, and complex systems: Differences between experts and novices. *Complexity, 6*(3), 41–49.

Jacobson, M., & Wilensky, U. (2006). Complex systems in education: Scientific and educational importance and implications for the learning sciences. *The Journal of the Learning Sciences, 15*(1), 11–34.

Jakobsson, E., & Working Group 1 Collaborators. *Complex systems: Why and what?* Retrieved February, 2006 from http://necsi.org/events/cxedk16/cxedk16_1.html.

Kaput, J., Bar-Yam, Y., Jacobson, M., Jakobsson, E., Lemke, J., Wilensky, U. and Working Group 1 Collaborators. *Two roles for complex systems in education: Mainstream content and means for understanding the education system itself.* Retrieved February, 2006 from http://necsi.org/events/cxedk16/cxedk16_0.html.

Lesh, R. (2006). Modeling students modeling abilities: The teaching and learning of complex systems in education. *The Journal of the Learning Sciences, 15* (1), 45–52.

Lesh, R., & Doerr, H. M. (Eds.). (2003). *Beyond constructivism: Models and modeling perspectives on mathematic problem solving, learning and teaching.* Mahwah, NJ: Lawrence Erlbaum.

Lesh, R., & English, L. D. (2005). Trends in the evolution of models & modeling perspectives on mathematical learning and problem solving. *International Reviews on Mathematical Education* (ZDM), 37, 6, 487–489.

Lesh, R., & Zawojewski, J. S. (2007). Problem solving and modeling. In F. K. Lester (Ed.), *Second Handbook of research on mathematics teaching and learning* (pp. 763–805). Greenwich, CT: Information Age Publishing.

Lesh, R., Cramer, K., Doerr, H. M., Post, T., & Zawojewski, J. S. (2003). Model development sequences. In R. Lesh & H. M. Doerr, (Eds.). (2003). *Beyond constructivism: Models and modeling perspectives on mathematic problem solving, learning and teaching* (pp. 35–58). Mahwah, NJ: Lawrence Erlbaum.

Romberg, T. A., Carpenter, T. P., & Kwako, J. (2005). Standards-based reform and teaching for understanding. In T. A. Romberg, T. P. Carpenter, & F. Dremock (Eds.), *Understanding mathematics and science matters.* Mahwah, NJ: Lawrence Erlbaum Associates.

Van den Heuvel-Panhuzen, M. (2003). The didactical use of models in realistic mathematics education: An example from a longitudinal trajectory on percentage. *Educational Studies in Mathematics, 54,* 9–35.

Yaneer B. (2000). NECSI: Application of complex systems: Sports and complex systems. Available at: http://www.necsi.org/guide/

Zawojewski, J. S, Lesh, R., & English, L. D. (2003). A models and modelling perspective on the role of small group learning. In R. A. Lesh & H. Doerr (Eds.), *Beyond constructivism: A models and modelling perspective on mathematics problem solving, learning, and teaching* (pp. 337–358). Mahwah, NJ: Lawrence Erlbaum.

ACKNOWLEDGEMENTS

The data reported here are from a study funded by a grant from the Australian Research Council. Any claims in this article are mine and do not necessarily represent the position of the Council. The support given by my research assistant, Jo Macri, is gratefully acknowledged.

APPENDIX A

Part A: Christmas Dinner at the Thompson's

John and Jan Thompson own and operate a convenience food store called the 'Food Owl.' Their shop is open seven days a week from 7 am to 7 pm. Both John and Jan work long hours in the store and very rarely get a day off.

Christmas is fast approaching and John and Jan Thompson are even busier than normal. People are coming into their shop to stock up on Christmas foods, decorations, small gifts, and holiday treats. Whilst other families are busy preparing for the festive season the Thompson's are busy *working*.

The Thompson's children, Dan, Sophie and Eva want to do something special for their parents this Christmas. They would like to make their parents limited holiday time as relaxing and enjoyable as they can.

Last year the children cooked a special dinner for their parents' Wedding Anniversary and they have decided to plan another celebration. This time they want to decorate the house, prepare, cook, and serve Christmas dinner for their Mum and Dad.

From their past experience Dan, Sophie and Eva know that planning a dinner party requires organisation and a well-planned schedule. Sophie said, "Once we do our schedule, the most overwhelming chore is done. Decorating, preparing and cooking the dinner are easy, but if the schedule isn't correct, then the whole dinner will never be ready!"

When the children organised their parents' Wedding Anniversary dinner they created an activity list and a schedule for preparing and cooking the meal.

Some of their jobs included:

ACTIVITY LIST

Menu and Instructions	
Chicken and sauce	
Preparation – chicken	30 min
Preparation – sauce	20 min
Cooking time	1 hour
Potatoes	
Wash, dry and cut up	15 min
Bake	50 min
Salad and dressing	
Wash and cut all ingredients and make dressing.	
Refrigerate until dinner time	20 min
Rolls	
Preparation	5 min
Warming time	10 min
Cheesecake dessert	
Preparation	30 min
Cooking time	1 hour

Other Tasks	
Set table	15 min
Organise serving dishes and utensils	15 min
Clean up	
After each cooking part	10–15 min
Dishing up and serving	15 min
Final clean up	
After serving	40 min

<table>
<tr><th colspan="10">PREPARATION AND COOKING SCHEDULE</th></tr>
<tr><td>Dan</td><td colspan="2">Prepare the chicken</td><td>Clean up from chicken</td><td colspan="2">Prepare cream sauce for chicken</td><td>Clean up kitchen</td><td colspan="2">Watch stove and oven</td><td>Dish food up and serve</td></tr>
<tr><td>Sophie</td><td colspan="2">Make the dessert</td><td>Clean up from dessert</td><td colspan="2">Wash salad, make dressing</td><td>Clean up from salad</td><td>Toss salad</td><td></td><td></td></tr>
<tr><td>Eva</td><td>Wash potatoes</td><td>Set table</td><td colspan="2">Whip cream for dessert</td><td>Clean up kitchen</td><td>Prepare serving dishes</td><td>Get rolls ready</td><td></td><td></td></tr>
</table>

Unfortunately, the schedule they made did not work well. They found that they stumbled around the kitchen wanting to use the same equipment at the same time. For example, Dan wanted to cook the chicken at the same time both Sophie and Eva wanted to use the oven!

The Thompson children realised that they didn't think of all the things they needed to include in their schedule.

PART B: What's Cooking at the Thompson's?

The children have already decided the menu for their Christmas Dinner:

- Before dinner nibblies (cheese, dip carrot sticks and water crackers)
- Baked turkey, roast vegetables and steamed vegetables
- Pavlova, ice-cream and fresh strawberries

Dan, Sophie, and Eva know that their parents will be home from work on Christmas Eve at 6:30 pm. All the children are available to begin preparing the Christmas celebration at 2 pm. They have four and one half hours to get everything ready! The three children hope that the Christmas celebration they plan for their parents will be successful and a wonderful surprise for their hardworking parents. All they have to do now is organise a schedule that works better than the wedding anniversary schedule.

Here are some of the things they need to consider:

- how long the turkey will take to cook
- what other items can cook in the oven with the turkey
- when to decorate and set the table
- when to make the pavlova and how long it will take
- how often they need to clean in between the cooking
- how much bench space they have for food preparation
- what food needs to be ready first
- who will use the equipment and when

They also need to consider who will be responsible for what jobs!

CHRISTMAS DINNER

Items for dinner	Preparation time and instructions	Cooking time
Turkey	20 min to stuff and prepare	3 hours oven time
Stuffing	3 min—to be done first	Cooks with the turkey
Carrots, beans and broccoli	Wash and cut—15 min	15 min steaming on the stove top
Roast potatoes and pumpkin	Wash and cut—15 min	Roast in oven—45 min
Salad	Wash and cut all ingredients—15 min Make dressing—10 min Toss salad—5 min	Chill in fridge for approx. 30 min
Gravy	Mixing—5 min	Heating on stove top—10 min
Bread	Cut and put on baking tray—5 min	Warm in oven—10 min
Pavlova	Separate eggs and whip egg whites—25 min	Bake pavlova in oven—1 hour
Decorate pavlova	Whip cream—10 min Wash and cut strawberries—10 min	Prepare pavlova for serving—10 min
Items to be ready at 6:30 pm		
Punch	Cut fruit mix juice and soft drink—10 min	Chill for approx. 2 hours
Cheese and crackers	Cut cheese and put out crackers—10 min	Do this as guests arrive home
Vegetables for dip platter	Wash and cut carrots into sticks—10 min	
Dip	Mix all ingredients—10 min	Chill for 1 hour

OTHER TASKS WHICH NEED TO BE COMPLETED

Task	Time
Clean house ready for celebration	30 min
Decorate house	45 min
Iron tablecloth and napkins	15 min
Set table	20 min
Clean ups (do after each food item e.g., stuffing, salad, pavlova and vegetables)	About 10–15 min each time

Part C: Your Challenge

Dan, Sophie, and Eva are making Christmas dinner for their parents. From previous experience they know they need a well planned schedule that they can follow. The schedule needs to factor in all the jobs they need to do and the times they need to allow for each job. The schedule should also help them determine which job should be completed first in order to have all the food ready at the correct time.

In the kitchen, there are two kitchen benches to work on, a double sink, a microwave oven, and a stove with four top burners and an oven. The oven is large enough to fit the turkey and one other item at the same time.

The Thompson children are expecting their parents home at 6:30 pm. They want to serve nibblies first, the turkey as main course and then the pavlova for dessert.

Dan, Sophie, and Eva need your help!! They have so many tasks to complete to be ready to surprise their parents, and they need a reliable schedule. Can you help them do two things?

1. Make a preparation and cooking schedule. Chart what each person will do and when, including the use of the kitchen equipment.
2. Write an explanation of how you developed the schedule. They plan to have other surprise celebrations for their parents and want to use your explanation as a guide for making future schedules.

APPENDIX B

Schedule Produced by the "Marvellous Mathematicians"

Serving Menu

6.30—cheese and crackers with vegetables and dip. Punch is also served at 6.30. 7 o'clock—warm bread. 7.15—main course, which includes stuffed turkey, steamed vegetables, roast pumpkin and potato. 8 o'clock—fresh salad with dressing. 8.30—pavlova with whipped cream and fresh strawberries.

Dan's Schedule

Stuff turkey at 3.40. 3.45—prepare turkey—put turkey in oven. 4.20—wash and cut carrots, bean, broccoli, potatoes and pumpkin. 4.25—steam carrots, bean and broccoli on stove. 4.50—steam vegetables, off stove.

4.45—mix gravy and put on stove. 5.25—take roast vegetables out of the oven. 7.05—take turkey out of the oven.

Sophie's Schedule

At 2.45—wash and cut ingredients for salad. 3 o'clock—make dressing and toss salad. At 3.30 cut and put bread on baking tray. 3.35—warm bread in oven. 4 o'clock—separate eggs and whip egg whites for pavlova. 7 o'clock—bake pavlova in oven. 7.15—whip cream wash and cut strawberries. 8 o'clock—take pavlova out of oven. 8.15—prepare pavlova for serving.

Eva's Schedule

At 4.20—cut fruit, mix juice and soft drink. 4.30—chill drinks for about 2 hours. 4.45—wash and cut carrots into sticks. 5.10—mix ingredients for dip and chill. 6.30—take out drinks and dip. 6.30—cut cheese and put out crackers.

All: Extra Activities

Eva—punch clean up, cheese clean, carrot, broccoli, clean up, set table, dip clean, cleaning.

Sophie—salad clean up, bread clean up, pavlova clean up, decorate house.

Dan—clean up turkey, clean up vegetables, clean up gravy, clean house and iron tablecloth.

CHAPTER 14

ATTITUDE TOWARD MATHEMATICS

Overcoming the Positive/Negative Dichotomy

Rosetta Zan and Pietro Di Martino
Dipartimento di Matematica, Pisa, Italy

INTRODUCTION

This contribution deals with "attitude toward mathematics," a construct which plays an important role in mathematics education. Our interest in research on attitude dates back to several years ago.

In previous work (Di Martino & Zan, 2001, 2002, 2003; Zan & Di Martino, 2003) we underlined the lack of theoretical clarity that characterizes research on attitude and the inadequacy of most measurement instruments; we analyzed the definitions of attitude that either explicitly or implicitly researchers most frequently refer to; we discussed the negative/positive dichotomy; we found a relationship between the risk of circularity in research on attitude and the researcher's implicit beliefs.

The Montana Mathematics Enthusiast, pages 197–214

In the following we will briefly summarize some results of these studies, in order to introduce recent findings from an Italian Project about attitude, aimed at investigating the phenomenon of "negative attitude towards mathematics."

The Lack of Theoretical Clarity in Research on Attitude

Research on attitude has a long history in mathematics education. The construct finds its origin in the field of social psychology (Allport, 1935), in connection with the problem of foreseeing individuals' choices in contexts like voting, buying goods, etc. Research develops more toward the formulation of measuring instruments than toward the theoretical definition of the construct, producing instruments that have given theoretical and methodological contributions of great importance (such as those by Thurstone and Likert).

In the field of mathematics education, research on attitude has been motivated by the belief that "something called 'attitude' plays a crucial role in learning mathematics" (Neale, 1969), but the goal of highlighting a *connection* between a positive attitude and achievement has not been reached. Ma & Kishor (1997), after analyzing the correlation of attitude/achievement in 113 classical studies, underline that this correlation is not statistically significant: they explain this to be caused by the inappropriateness of the observing instruments that were used (in our opinion not only related to attitude, but also to achievement).

The attitude construct gains renewed popularity with the re-evaluation of affect in the learning of mathematics: in the classification of Mc Leod (1992) it is considered together with *beliefs* and *emotion* one of the constructs that constitute the affective domain (De Bellis & Goldin, 1999, propose *values* as a fourth construct).

Even if the meaning of the various terms is not always agreed upon, or even made explicit (Hart, 1989; Pajares, 1992) there is consensus on the fact that emotions and beliefs deeply interact: as regards attitude, an emotional component is generally explicitly recognised in the construct, often together with a cognitive component, mainly identified with beliefs.

Most researchers have underlined the need of some theory for research on affect, in order to better clarify connections among the various components, and their interaction with cognitive factors in mathematics education (Mc Leod, 1992). Recently the need for a theory for affect has been given several kinds of answers, differing for both the particular construct explicitly or implicitly chosen as 'starting' point (for example emotions or beliefs), and the different focus.[1]

Research on attitude has been judged to be particularly contradictory and confusing, due to the fact that it has given more emphasis to creating measurement instruments rather than elaborating on a theoretical framework (Kulm, 1980; McLeod, 1992; Ruffell, Mason & Allen, 1998): in actual fact, the work providing the most consistent information focused on the description of the *differences* between groups of people, usually males and females (Fennema, 1989).

Defining Attitude toward Mathematics

The lack of theoretical framework that characterizes research on attitude toward mathematics is partially shown by the fact that a large portion of studies about attitude do not provide a clear definition of the construct itself: attitude tends rather to be defined implicitly and a posteriori through the instruments used to measure it (Leder, 1985; Daskalogianni & Simpson, 2000).

When a definition is explicitly given, or can be inferred, it mainly refers to one of the three following types:

1. A 'simple' definition of attitude, that describes it as the positive or negative degree of affect associated with a certain subject. According to this point of view the attitude toward mathematics is just a positive or negative emotional disposition toward mathematics (McLeod, 1992; Haladyna, Shaughnessy J. & Shaughnessy M., 1983).
2. A multidimensional definition, which recognizes three components in the attitude: emotional response, beliefs regarding the subject, behaviour related to the subject. From this point of view, an individual's attitude toward mathematics is defined in a more complex way by the emotions that he/she associates with mathematics (which, however, have a positive or negative value), by the individual's beliefs towards mathematics, and by how he/she behaves (Hart, 1989).
3. A bi-dimensional definition, in which behaviours do not appear explicitly (Daskalogianni & Simpson, 2000): attitude toward mathematics is therefore seen as the pattern of beliefs and emotions associated with mathematics.

Kulm (1980) suggests that "It is probably not possible to offer a definition of attitude toward mathematics that would be suitable for all situations, and even if one were agreed on, it would probably be too general to be useful" (p. 358). In this way, the definition of attitude assumes the role of a '*working definition*' (Daskalogianni & Simpson, 2000). This position views the attitude construct as functional to the researcher's self-posed problems:

in these terms we consider it to be useful in the context of mathematics education, as long as it is not simply borrowed from the context in which it appears, i.e. social psychology, but is rather outlined as an instrument capable of taking into account peculiar problems in mathematics education. This is in line with the position of Ruffell, Mason and Allen (1998), who see attitude as an observer's construct.

What Does Positive or Negative Attitude Really Mean?

In actual fact the term attitude is used in both practice and research together with the adjectives *positive/ negative.*

This dichotomy between positive/negative attitude pervades mathematics education research, both implicitly and explicitly. For example, classic studies regarding the relationship between attitude and achievement in fact investigate the correlation between *positive* attitude and success. In the same way studies aiming to change attitude actually end up in setting the objective of transforming a negative attitude into a positive one.

The definition of positive or negative attitude toward mathematics clearly depends on the definition of attitude itself.

According to the simple definition, it is clear what a positive or a negative attitude is: a *positive attitude* is a positive emotional disposition toward the subject; a *negative attitude* is a negative emotional disposition toward the subject.

If we choose the bidimensional (or multidimensional) definition, it is not clear what a positive attitude should mean, but referring only to the emotional dimension is reductive, since we have to take into account the two (three) dimensions, i.e., emotions, beliefs, (behaviours) and their interaction.

What actually happens is that in most studies the choice of a definition for attitude, and consequently a characterization of positive/negative attitude, not only is not explicitly made: often it is not made at all, and the assessment/ measurement instruments used by the researcher implicitly end up by continuously wavering between various definitions within a single study.

The characterization of an individual's attitude as positive/negative is in most cases simply the result of a process of measurement, performed through instruments such as the Thurstone or Likert attitude-scales or the semantic differential technique.[2] This process ends up in a score—attached to an individual's attitude—obtained by summing points relating to the single items. The choice of scores to be assigned to the items naturally leads to a positive/negative evaluation of each one.

Since in most questionnaires used to assess attitude the items range from those related to emotions ("I like mathematics") to those related to beliefs

("Mathematics is useful"), to those related to behaviour ("I always do my homework in maths"), an answer can be characterised as positive by referring to different meanings of the word *positive* itself. More precisely, this meaning varies depending on whether positive refers to emotions, beliefs, or behaviour:

1. When it refers to an emotion, positive normally means "perceived as pleasurable." So, anxiety when confronting a problem is seen as negative, while pleasure in doing mathematics is evaluated as positive.
2. When it refers to beliefs, positive is generally used with the meaning "shared by the experts."
3. When it refers to behaviour, positive generally means "successful." In the school context, a successful behaviour is generally identified with high achievement: this naturally poses the problem of how to assess achievement (Middleton & Spanias, 1999).

In actual fact the three meanings overlap. For example, in the case of beliefs, sometimes positive means that it is supposed to elicit a positive emotion. A typical case is represented by the belief "Mathematics is useful," which is also used in questionnaires aimed at measuring just the emotional dimension of attitude (i.e., the simple definition of attitude: see Haladyna, Shaughnessy & Shaughnessy, 1983). But often positive referred to a belief means that it is supposed to be related to a positive behaviour, i.e., to a successful behaviour. Sometimes the latter meaning is also used for emotions, implicitly admitting that a positive emotion toward mathematics, being pleasurable, is necessarily associated with a positive behaviour in mathematics. On the contrary several studies (Evans, 2000) suggest the possibility that for certain subjects an optimal level of anxiety exists: above this level, but also below it, performance is reduced. The problem is that, generally differences between the various meanings are rarely made explicit: in this way, an a priori assumption is often made as to what should in effect be the result of an investigation, for example, that a belief which is positive because it is shared by experts, is associated with a positive behaviour in that it is successful. This continuous sliding between the researcher's assumptions and the desired result of the investigation enhances the risk of *circularity* in research on attitude, a risk that Lester (2002) more generally pointed out for research on affect.

Depending on the criteria used to evaluate an attitude, different results may be obtained: for example, an attitude can be evaluated positive as regards the emotional dimension, but negative regarding the cognitive dimension, or vice versa.[3] The problem is only apparently overcome when the algebraic sum of the two components results in a single evaluation. Furthermore, as we have observed, beliefs are often used to assess the significance of the emotional dimension, or evaluated according to their behavioural consequences, and this increases ambiguity.

The differences in the use of the adjective *positive* not only imply different choices of assessment/measurement instruments: it also triggers a different formulation of the research problem to be dealt with. For example, the problem of identifying how to push a positive attitude, typically encountered in this field of research, requires a completely different approach depending on whether the positive attitude refers only to the emotional component or it refers to a particular pattern of beliefs and emotions, to be assumed as a model.

AN ITALIAN PROJECT ABOUT ATTITUDE

The points made above about *the need for a theoretical framework for affect*, together with the importance of *linking theory and practice*, have been fundamental issues of an Italian Project about attitude. Several Italian researchers participated in the project, named "Negative attitude towards mathematics: Analysis of an alarming phenomenon for culture in the new millennium."[4] The project's main objective was to investigate the phenomenon of "negative attitude towards mathematics," viewed as something that is connected to the learning of the discipline, but that also affects various aspects of the social context: the refusal of many students to enrol in scientific undergraduate courses due to the presence of exams in mathematics, a worrying mathematical illiteracy, an explicit and generalized refusal to apply rationality characterising scientific thinking, or, vice versa, a tendency to uncritically accept models that are only apparently rational.

One of the project's sub-goals was to provide an operative definition of positive or negative attitude toward mathematics, capable of giving teachers and researchers theoretical tools to observe and interpret some difficulties students meet in mathematics and possibly suggest ways to overcome these difficulties.

To reach this goal a theoretical reflection was needed, continuously and deeply linked to investigations about both origins of a negative emotional disposition towards mathematics and factors that influence its development. More precisely:

- a longitudinal investigation on a sample of students, covering the three-year duration of the project;
- collateral investigations performed on other subjects and involving teachers, students' family members, adults in general, and professional mathematicians.

The methodology entailed an integrated method approach:

- the use of questionnaires, diaries and interviews for the observation of teachers and adults;
- class observation, questionnaires, structured and semi-structured interviews, conversations, essays, etc. for monitoring students over the three years.

In the following, we will briefly present some results of this Project, that in our opinion can give a contribution to the theoretical issues discussed earlier.

INVESTIGATION ON TEACHERS' USE OF THE "NEGATIVE ATTITUDE TOWARDS MATHEMATICS" CONSTRUCT

Within the Project, one of the activities conceived to favour a link between theory and practice has been an investigation carried out with teachers, aimed to reach two objectives:

1. to see whether in their practice teachers use the construct of negative attitude when they diagnose difficulty;
2. if this is the case, to see *how* they use it, investigating:
    - what type of definition they make reference to (in particular, whether they use the 'simple' definition which sees attitude simply as an emotional disposition towards mathematics);
    - if and how the diagnosis of negative attitude constitutes an instrument for intervening in a more targeted way on recognised difficulties.

These aspects were investigated through a questionnaire, administered to 146 teachers from various school levels: 29 from primary school, 50 from middle school, and 67 from high school (Polo & Zan, 2005). The questionnaire was specifically constructed to find out teachers' beliefs towards the negative attitude pupils can have towards mathematics, and contained 6 multiple choice questions and 6 open ended questions (see Fig. 14.1). The multiple choice questions aimed at discovering whether and how frequently teachers use the attitude construct in the diagnosis of difficulty, and if they consider changing a negative attitude at the end of high school a possible thing. The open ended questions intended to investigate what idea teachers have of negative attitude, and what indicators they use as reference.

School: ______________________________

☐ M ☐ F Age: __________ Date: ______________

1. Do you ever find yourself attributing a pupil's difficulties with mathematics to his/her *attitude* towards the subject?

   ☐ Yes ☐ No

2. If yes, is this a frequent diagnosis or have you only seen it a few times?

   ☐ practically never ☐ sometimes ☐ rarely ☐ often ☐ nearly always

3. What do you mean by *negative attitude* towards mathematics?

4. What demonstrates to you that a student has a negative attitude towards

5. Do you think it is possible to modify the attitude of a pupil at the end of high school?

   ☐ yes ☐ only to a certain extent ☐ maybe ☐ no ☐ don't know

6. If yes, *how*? If no, *why*?

7. Have you ever set yourself the specific objective of changing the attitude of one of your pupils?

   ☐ Yes ☐ No

8. If yes, how did you attempt to achieve this? What were the results?

9. Up to now we have only referred to a single student. Have you ever seen a negative attitude towards mathematics in a whole class?

   ☐ Yes ☐ No

10 How did you recognise this negative attitude?

11. If you answered yes to question 9, in this case did you explicitly set yourself the objective of changing the attitude of the class?

   ☐ Yes ☐ No

12. If yes, how did you try to reach this objective? What was the result?

Figure 14.1 The questionnaire

The answers to Question 1 highlight the wide use of the term 'attitude' in relation to the diagnosis of a pupil's difficulty in mathematics: 85,6% of the sample (125 out of 146) in fact gave a positive answer to the first question. A comparison of the answers given by teachers from the various school levels shows a peak in the positive answers at middle school level (47 teachers out of 50, equal to 94%), with respect to the homogenous answers provided by primary and high school teachers (24 out of 29, equal to 82,8%, and 54 out of 67, equal to 80,6% respectively). The answers to question 2 confirm the frequency in the use of the construct: 70 teachers out of 125 (equal to 56%) answer that they attribute a pupil's difficulties with mathematics to his/her attitude toward the subject sometimes/often/nearly always. A comparison between the various school levels highlights that the number of responses of that kind (sometimes/often/nearly always) increases with schooling level: 8 questionnaires out of 21 (equal to 38,1%) at elementary school level, 28 out of 40 (70%) at middle school level, and 34 out of 45 at high school level (75,6%).

The answers to the open ended questions give additional significant information:

- most teachers seem to have, even implicitly, the multidimensional idea of attitude, rather than the simple one: they refer to students' beliefs about mathematics (in particular the belief that being quick in mathematics is very important, that mathematics is made of mechanistic rules, that mathematics is useless, difficult,...), and to self-efficacy beliefs (students' claims such as "I am not inclined to mathematics," "Since primary school I always failed in mathematics");
- the teacher's apparently natural use of the 'negative attitude' multidimensional construct goes together with a lack of a clear distinction between the definition of attitude and the identification of indicators, thus making the definition itself not operative;
- the causes of a negative attitude are generally ascribed to *students'* characteristics and behaviours, thus hiding the teacher's responsibility in building a view of mathematics that elicits refusal, in the lack of interest and effort by students, in the image of the self that students construct;
- but most of all the diagnosis of negative attitude, referred to a single student, seems to be the final result of the teacher's interpretative process of the student's failure, rather than the starting point of a remedial action.

These results suggest that the attitude construct—as used by teachers—although rich enough to deal with the complexity of learning mathematics, does not seem to have the characteristics of a theoretical instrument capable of directing their work (particularly in that of helping students recover from difficulties): teachers rather seem to recognise a situation that is difficult to manage and modify.

In the end, this study highlights the importance of producing a definition of attitude toward mathematics, capable of making this construct a theoretical tool to direct teachers' observation, interpretation, remedial actions.

INVESTIGATING ATTITUDE THROUGH ESSAYS

In our Project the aim of giving a 'suitable' definition of attitude toward mathematics has been attained by observing students and collecting information through questionnaires, structured and semi-structured interviews, conversations, essays, etc..

One of the instruments used has been an essay on mathematics: "Me and mathematics: my relationship with maths up to now".

With this instrument we meant to get over the normative approach that characterises most research on attitude, and that we consider one of the reasons underlining both the lack of theoretical clarity and the difficulties encountered in getting significant results. Thus we adopted an interpretative approach, investigating students' relationship with mathematics from the bottom and trying to spot in their descriptions the dimensions involved.[5] In other words, we did not assume an *a priori* definition of attitude toward mathematics, but we rather referred to the more general relationship with mathematics in posing questions that underpin studies on attitude in mathematics education:

- How does attitude toward mathematics evolve during a subject's school experience?
- How can we explain the spoiling of attitude toward mathematics from elementary school to high school?
- What are the variables that influence attitude toward mathematics? Which of them are controllable?
- Is it possible to modify a subject's attitude toward mathematics? How?

We ask the subjects to tell their own story with mathematics through an autobiographical essay: in doing this we are convinced that pupils will tend to explicitly evoke those events and remarks about their past that they deem important "here and now" and they will also tend to paste fragments, introducing some causal links, not in a logical perspective but rather in a social, ethical and psychological one (Bruner, 1990).

We assume that, in order to describe the kind of relationship an individual has with mathematics, this "pasting" process—typical of autobiographical narratives—is more important than an "objective" report of one's experience with the discipline at school. In other words, we agree with Bruner's (1990) claim that it is not important whether the story told is actually "contradictory" or "likely": we are rather interested in what the individual thinks he/she has done, the reasons underlying these actions, the type of situations he/she believed to be into and so on.

The essay was proposed at all school levels (for primary school the title was abbreviated in "*Me and mathematics*"): we explicitly asked that the essay be not proposed by the mathematics teacher. We collected a huge amount of materials, coming from different geographic areas and schools, but our sample cannot be viewed as a representative one, since it is based on both schools and teachers' spontaneous will to participate.

Overall we collected 1304 essays: 741 from primary school, 256 from middle school, and 306 from high school.

The analysis was carried out, according to an interpretative approach, trying to *understand* how students interpret their own experiences with mathematics, rather than to *explain* their mathematical path in terms of cause/effect. Final outcome of this analytical process is expected to be the construction of a set of categories, properties, relationships: a *grounded theory* (Glaser & Strauss, 1967), i.e., a theory based on collected data, the construction of which requires a continuous back and forth between the different research phases. In our case the essays were read in the light of both pre-existing categories (for instance liking and disliking mathematics) and in a free way, trying to identify meaningful categories *a posteriori*.

Although our type of sample requires a careful evaluation before any generalisation of data be made, significant points for analysis also come from quantitative-type reflections. Due to the huge amount of essays, we decided to use a specific piece of software for textual analysis, called T-Lab, to carry out this type of analysis.

SOME RESULTS

As we mentioned earlier, we will only present here some results that can bring a significant contribution to the theoretical problems we dealt with so far.

Reading the essays, we identified three *core themes*, and precisely:

- the emotional disposition towards mathematics, concisely expressed with "I like/dislike mathematics";
- the perception of being /not being able to succeed in mathematics, concisely expressed with "I can do it/I can't do it";
- the vision of mathematics, concisely expressed with "mathematics is…."

The hypothesis suggested by a reading of the essays is confirmed by a quantitative analysis carried out through T-LAB: the most frequent expression in the about 1,300 essays collected is "I like" (in the different forms: "I like/I don't like/I used to like..."), followed by "I can do it/I can't do it" and "mathematics is...."

Sometimes an essay develops around one of these three themes: more often, it makes reference to all the themes, although it is centred on one of them (which is therefore called the *core theme* of the essay). It is meaningful from both a theoretical and an educational perspective that these three themes are deeply connected: this clearly comes out from a reading of the essays and will be the basis for some remarks in the following.

The most frequent connection is associated with the word *because*: starting from the most recurrent theme "I like/dislike," it is a motivation ("I like /dislike *because*...") that leads to one of the other two themes: the vision of mathematics or the perception of being/not being able to succeed. The vision of mathematics is also brought in when the core theme is the perception of being/not being able to succeed (I can/can't do it), once again through the underlying reasons: "I can/can't do it *because*...."

Let us examine these connections.

I Like/Dislike → Vision of Mathematics

This path highlights the fact that different emotional reactions (like/ dislike) mainly link to different visions of mathematics

A widely spread distinction between these two visions of mathematics is that drawn by Skemp (1976) between one kind of mathematics characterised by "rules without reasons"—named *instrumental*—juxtaposed to one vision of mathematics as *relational*, in which understanding means "knowing both what to do and why."

Often "I like mathematics" is associated with a relational vision, and "I dislike mathematics" to an instrumental one.[6]

> I never liked to learn things by heart (except for some formulae) and this subject, together with Physics, gives me a chance to think and discuss. I like it, because it is a subject which needs reasoning. [3H.16]

> I don't like it because there are many rules to make a tiny little operation: you must divide one number by the other one, take away the number you had before and so on. Moreover, if you forget a rule you run into troubles! [1M.16]

Sometimes the instrumental vision of mathematics is associated with a positive emotional reaction:

> As I was growing up I got more and more fond of mathematics, because I starter to do expressions (the things I prefer in this subject); and so, after seeing that I was starting to get them right, then mathematics or, better say algebra, became my passion. [2H.74]

In the end, sometimes one same vision of mathematics can be associated with different emotional reactions. This is the case for the characterisation of mathematics as rational:

> It is fascinating because it is not an opinion, it is a rational subject (like my own character), which needs no interpretation; ... it is so. [5H.4]

> (...) this does not mean that I like mathematics, actually I completely hate it, simply because it is a subject I feel really far from me. When you have to solve an equation you don't need to be creative, to interpret or say what you feel; mathematics is empty of feelings, just think of the well-known saying: "mathematics is not an opinion". [5H.1]

To conclude, these data stress the fact that the two dimensions—vision of mathematics and like/dislike—are mutually independent.

This independence is strongly expressed in characterising mathematics as useful/useless, and easy/difficult. These dichotomies are particularly interesting to our discussion, since the items "Mathematics is useful" (with a positive score to an affirmative answer) and "Mathematics is difficult" (with a negative score to an affirmative answer) are frequently used in questionnaires about attitude.

Reading the essays we have found all four combinations of like/dislike and easy/difficult:

> I like mathematics because...

> ... *it is easy*: Since primary, I always found mathematics rather simple and easy to be understood. And maybe this is the reason why it has always been one of my favourite subjects. [2H.73]

> ... *it is difficult*: The story of mathematics in my life started off with logic: sets of fruits, difference between "and" and "or", opposites (like red and not red) [...]. As time was going by, as fruits disappeared and notions and difficulties were increasing, mathematics became more and more interesting and involving and my judgement changed completely. [5H.16]

> I don't like mathematics because...

> ... *it is easy*: In the beginning mathematics was nasty because it was too easy: 3 + 1, 5 + 5,... [4P.115]

> ... *it is difficult*: I didn't like mathematics much because I saw it was difficult and I did not manage I gave up quickly. [1H.36]

As regards the useful/useless dichotomy, essays very often show that even when mathematics is perceived as useful, this perception is not necessarily associated with a positive emotional disposition:

> Although it is a useful subject, I don't like it. [1H.3]

I Like/Dislike → I Can/Can't Do It

The connection beween the theme "I like" and the theme linked to self-efficacy, expressed through 'I can / can't do it' comes out so strong from the essays that sometimes the expressions "I like" ("I dislike") and "I can do it" ("I can't do it") are used as synonyms:

> Since primary school, I remember when the teacher asked us to number by 2, 3, 6, 9 up to 800, 900 ... I used to hate it. Then I changed school and I started to hate it even more because of the expressions. Let's not talk about middle school I changed 4 teachers in the 3 school years and therefore if I didn't understand anything before, now I really understand zero. [1H.3]

In the greatest part of the essays in which the dimensions "I like/dislike" and "I can/I can't do it" are connected, this connection is *"I like it because I can do it"* and *"I dislike it because I can't do it."* In some rare cases, we found the combinations *"I like it although I can't do it"* and *"I can do it but I dislike it."*

From both an educational and a theoretical perspective, the most interesting outcome of the reading of the essays is that success in mathematics has many deeply different meanings. In some essays succeeding is identified with school success, i.e., with getting good marks, and thus it is up to the teacher to acknowledge one's success. In some other cases, succeeding is identified with understanding. In the latter case things are come complicated: sometimes understanding is used with an *instrumental* meaning, and it is identified with knowing the rules and being able to apply them correctly; in other cases a *relational*-type understanding appears, referring to one's awareness of why the rules work and how they are linked to one another.

I Can/Can't Do It → Vision of Mathematics

When perception of being/not being able to succeed ("I can/can't do it") is the core theme of the essay often the pupil talks about the reasons underlying his/her success or failure explicitly: these are the so called *causal attributions* for success and failure (Weiner, 1974). These attributions often allow us to recognise not only the pupil's beliefs about him/herself but also his/her vision of mathematics, sometimes through the so-called *theories of success*, i.e., the theories a pupil may have about success in mathematics.

Once again a wide range of theories emerges, corresponding to a wide range of ways of viewing mathematics: for instance, theories of success or attributions that centre on the important role played by memory, suggest an instrumental vision of mathematics, whereas theories of success and attributions focusing on the need to understand what one is doing, suggest a relational vision of mathematics:

> Up to middle school I have always succeeded in mathematics, because I always understood the reasoning behind it. [1H.15]

> It is not that I don't understand maths, it's that I make a mess, because it's full of rules and theorems and it's almost impossible to remember each of them, moreover when I'm finally at ease with a topic, it seems they make it on purpose: we carry on with the planned contents and I am cheated. [2H.20]

CONCLUSIONS

The results presented here confirm some issues we discussed in the beginning.

The choice of using agreement on certain beliefs to deduce a positive emotional reaction is questioned by the fact that different emotional reactions (like/dislike) can be associated with one single vision of mathematics. This is particularly true for two widely used items: "Mathematics is useful," generally viewed as positive, and "Mathematics is difficult," generally viewed as negative.

Analysing the collected essays we observed that the majority of them develop around two recurrent and generally interconnected core themes: *I like/dislike* and *I can/can't do it*. These themes are intertwined in the author's *vision of mathematics*: in particular, one single emotional reaction can be associated with conflicting visions of mathematics, and possibly conflicting emotional reactions can be associated with different visions of mathematics.

The essays describe pupils' relationship with mathematics as referred to three dimensions: like/dislike, perception of being/not being able to succeed, vision of mathematics. In the sample we examined these three dimensions are combined in many different ways.

These results suggest that for a description of a pupil's attitude towards mathematics it is not enough to highlight his/her (positive/negative) emotional disposition towards the discipline: it is necessary to point out what vision of mathematics and what self-efficacy beliefs this emotional disposition is associated with.

If we start from this more complex definition for attitude, a problem arises about what should be meant by positive or negative attitude. Referring to the only emotional component seems to have limitations from both a theoretical and an educational perspective: this would lead us to define as positive the attitude of a pupil who views mathematics as a discipline made of rules to memorise and apply rigidly, only because he likes the subject itself.

It is thus possible to introduce different types of negative attitudes, different profiles of attitude, depending on the component the adjective negative refers to. It might be referred to a distorted and epistemologically wrong vision of mathematics, and as such negative, as in the example discussed above. Rather, it might be referred to the *beliefs about the self* component, outlining as negative those beliefs about the self which are characterised by a scarce sense of self-efficacy ("I'm not able," "I can't make it," "I'm hopeless").

Some recurrent patterns can be traced in the diverse combinations emerging from the essays. For instance, in essays telling a story of difficulties or unease, an instrumental vision of mathematics often emerges, together with negative emotions, identification of success and achievement, theories of success that stress the role of memory, scarce sense of self-efficacy. The recurrence of these combinations make them significant from an educational point of view: the teacher is thus provided with hints on possible routes to be followed to enable pupils to *narrate a different story*.

After all, this way of seeing attitude towards mathematics—constructed as a grounded theory—may become a useful instrument for both teachers and researchers. The diagnosis of a negative attitude becomes a starting point for the teacher to design an intervention aimed at modifying the component(s) identified as negative for that pupil. As for researchers, this definition, as it has been constructed, provides a strong link with practice, prevents from falling into circularity and, in the end, allows them to overcome some of the critical points of research on attitude.

NOTES

1. See for example the Research Forum about "Affect in mathematics education: Exploring theoretical frameworks" at PME 2004 (Hannula, Evans, Philippou & Zan, 2004), and the PME Special Issue with the same title (Evans, Hannula, Zan & Brown Eds., 2006). As regard beliefs, a considerable work of reorganization and systematisation has been done in the book *Beliefs: A Hidden Variable in Mathematics Education?* (Leder, Pehkonen & Törner Eds., 2002).
2. This characterisation is not so frequent in qualitative studies, and when it is used, it is generally accompanied by a description of the factors (behavior, beliefs, emotions) from which it is obtained. In any case, the evaluation of a positive attitude brings us back to a positive evaluation of at least one of the components: emotions, beliefs, behaviour.
3. This is what Hannula (2002) observes, describing the evolution of the attitude of Rita, a lower secondary school student: he underlines that, using the term *attitude* in a traditional manner, *"in the beginning Rita had an 'attitude' that was negative and positive at the same time"* (p. 42).
4. Besides the authors, the researchers participating in the Project are Pier Luigi Ferrari, Fulvia Furinghetti, Donatella Iannece, Paolo Lorenzi, Nicolina Malara, Maria Mellone, Francesca Morselli, Maria Polo, Roberto Tortora.

5. In the field of affect the need for social and anthropological approaches, i.e., for studying affect in its natural contexts, is particularly stressed, motivating the use of non-traditional methods, such as narratives (see for example Ruffell, Mason & Allen, 1998; Hannula, 2003).
6. In the following excerpts the first number refers to the class level, the letter refers to the school level (Primary/Middle/High), the last number indicates the progressive numbering of the essay within the category.

REFERENCES

Allport, G.W. (1935). Attitudes. In C.A. Murchinson (Ed.) *A handbook of social psychology*. Worcester, Mass: Clark University Press.

Bruner, J. (1990). *Acts of Meaning*. Cambridge: Harvard University Press.

Daskalogianni, K. & Simpson, A. (2000). Towards a definition of attitude: the relationship between the affective and the cognitive in pre-university students. *Proceedings of PME 24*, vol.2, 217–224, Hiroshima, Japan.

DeBellis, V. & Goldin, G.A. (1999). Aspects of affect: mathematical intimacy, mathematical integrity. *Proceedings of PME 25*, vol.2, 249–256, Haifa, Israel.

Di Martino, P. & Zan, R. (2001). Attitude toward mathematics: some theoretical issues. *Proceedings of PME 25*, vol.3, 351–358, Utrecht, Netherlands.

Di Martino, P. & Zan, R. (2002). An attempt to describe a 'negative' attitude toward mathematics. *Proceedings of the MAVI-XI European Workshop*, 22–29, Pisa, Italy.

Di Martino, P.& Zan, R. (2003). What does "positive" attitude really mean? *Proceedings of the 27th Conference of the International Group for the Psychology of Mathematics Education*, vol. 4, 451-458, Honolulu, Hawai'i.

Evans, J. (2000). *Adults' Mathematical Thinking and Emotions*. London: Routledge Falmer .

Evans, J., Hannula, M., Zan, R., Brown, L. (Eds.) (2006). Affect in Mathematics Education—Exploring Theoretical Frameworks. *Educational Studies in Mathematics, Special Issue.*

Fennema, E. (1989). The Study of Affect and Mathematics: A Proposed Generic Model for Research. In Mc Leod & Adams (Eds.) *Affect and Mathematical Problem Solving* (pp. 205–219). New York: Springer Verlag.

Glaser, B.G. & Strauss, A.L. (1967). *The Discovery of Grounded Theory. Strategies for Qualitative Research*. Chicago: Aldine.

Haladyna, T., Shaughnessy, J., Shaughnessy, M. (1983). A causal analysis of attitude toward Mathematics. *Journal for Research in Mathematics Education*, 14 (1), 19–29.

Hannula, M. (2002). Attitude toward mathematics: emotions, expectations and values. *Educational Studies in Mathematics*, 49, 25–46.

Hannula, M. (2003). Affect towards mathematics; narratives with attitude. In M. A. Mariotti (Ed.), *Proceedings of the Third Conference of the European Society for Research in Mathematics*. [CD] Pisa, Italy.

Hannula, M., Evans, J., Philippou, G., Zan, R. (2004). Research Forum: Affect in mathematics education—exploring theoretical frameworks. *Proceedings of PME 28*, vol.1, 107–136, Bergen, Norway.

Hart, L. (1989). Describing the Affective Domain: Saying What We Mean. In Mc Leod & Adams (Eds.) *Affect and Mathematical Problem Solving* (pp. 37–45). New York: Springer Verlag.

Kulm, G. (1980). Research on Mathematics Attitude. In R.J. Shumway (Ed.), *Research in mathematics education* (pp. 356–387). Reston, VA: NCTM.

Leder, G. (1985). Measurement of attitude to mathematics. *For the Learning of Mathematics,* 34 (5), 18–21.

Leder, G., Pehkonen, E., Törner, G. (Eds.) (2002). *Beliefs: A Hidden Variable in Mathematics Education?* Dordrecht: Kluwer Academic Publishers.

Lester, F. K. Jr. (2002). Implications of Research on Students' Beliefs for Classroom Practice. In G. Leder, E. Pehkonen & G. Törner (Eds.), *Beliefs: A hidden variable in mathematics education?* (pp. 345–353). Dordrecht: Kluwer Academic Publishers.

McLeod, D. (1992). Research on affect in mathematics education: a reconceptualization. In D.Grows (Ed.), *Handbook of Research on Mathematics Teaching and Learning* (pp. 575–596). New York: McMillan Publishing Company.

Ma, X. & Kishor, N. (1997). Assessing the Relationship Between Attitude Toward Mathematics and Achievement in Mathematics: A Meta-Analysis. *Journal for Research in Mathematics Education,* 28 (1), 26–47.

Middleton, J.A. & Spanias, P.A. (1999). Motivation for Achievement in Mathematics: Findings, Generalizations, and Criticism of the Research. *Journal for Research in Mathematics Education,* 30, 65–88.

Neale, D. (1969). The role of attitudes in learning mathematics. *The Arithmetic teacher,* Dec. 1969, 631–641.

Pajares, F. (1992). Teachers' Beliefs and Educational Research: Cleaning Up a Messy Construct. *Review of Educational Research,* 62 (3), 307–332.

Polo, M. & Zan R. (2005). Teachers' use of the construct 'attitude'. Preliminary research findings. In M. Bosch (Ed.), *Proceedings of the Fourth* Conference of the European Society for Research in Mathematics. http://ermeweb.free.fr/CERME4/CERME4_WG2.pdf, 265–274.

Ruffell, M., Mason, J., Allen, B. (1998). Studying attitude to mathematics. *Educational Studies in Mathematics,* 35, 1–18.

Zan, R. & Di Martino, P. (2003). The role *of* affect in the research *on* affect: the case of 'attitude'. In M. A. Mariotti (Ed.), *Proceedings of the Third Conference of the European Society for Research in Mathematics.* [CD] Pisa, Italy.

CHAPTER 15

SUBJECT-RELATED EDUCATION MANAGEMENT

Course Concept and First Findings from Accompanying Research[1]

Konrad Krainer and Florian H. Müller
University of Klagenfurt, Austria

Abstract: The target group for the four-semester professional development course entitled "Subject-Related Education Management" (2006–2008) in Austria are secondary school teachers of German, mathematics and the natural sciences. They are expected to initiate and support processes in their respective federal states with the aim of critically appraising individual and interdisciplinary teaching methods for each subject, and to (further) develop new methods. The goal of accompanying research is to examine participants' development of competence. Initial findings show that there are no significant differences between teachers with regard to the self-assessment of interests and competences in relation to subject area, gender or school type. However, the extent to which teachers have experience in the organisation and implementation of teacher education makes a significant difference.

The Montana Mathematics Enthusiast, pages 215–230

THE IMPORTANCE OF SUBJECT-RELATED EDUCATION MANAGEMENT

The IMST Project[2] (cp. Krainer, 2007) aims at introducing and supporting innovations in mathematics, science and technology teaching at Austrian schools. In addition, some *structural measures* are established at the education system level.

This includes, for instance, setting up national competence centres for didactics in several subjects, regional networks and *subject-related education management.* Some schools and federal states already have certain teachers who take on specific tasks for one or more subjects. However, these arrangements are usually not formally anchored and are based—with a few exceptions—more on vague descriptions of function and task. Despite Austria's overseeable size, these arrangements differ significantly from one state to the next with regard to tasks, personnel selection, anchoring within the regional educational system as well as payment. The exchange among those concerned across Austria is rudimentary. The autonomy of schools (and educational regions) is usually understood and practiced as the autonomy of individuals. This is often supported and enhanced by a lack of agreement between and within educational institutions.

In contrast to this situation, subject-related and subject-connecting challenges require collective action and reflection. These stretch from an exchange of experience and mutual agreement of aims and performance requirements, determining main focus points with regard to content and methodology (e.g., justification for new subjects or the introduction of laboratory teaching) to describing the overall concept of a subject (incl. the question of a subject's contribution to education). Achieving greater significance and consideration for output orientation (educational standards, etc.) will require (in addition to the continued consideration of input questions) in particular processes and structures to support the orientation.

The development of internal structures and the co-operation and coordination between schools will in future play a central role in achieving and improving sustainable development in Austria's education system. Subject-related middle management is needed at school and regional level to support and nurture school development whilst remaining close to teaching methodology and the subjects themselves. In the long term, these functions should only be fulfilled by persons with appropriate education and qualification. It would also represent an important contribution to the further professionalization of the teaching profession. Subject-related education managers would also make a contribution towards the *improvement of communication and the exchange of knowledge* between educational practice, education authorities and educational science, thus creating a significant prerequisite for the sustainable development of the education system.

In a first step, the Federal Ministry for Education has assigned the development of a university course entitled "Subject-related Education Management" (2006–2008) that will initially apply at a regional level. The target group of the four-semester course consists of lower secondary school teachers (secondary modern in Great Britain, junior high school in the USA) teaching German, mathematics or natural sciences (biology, chemistry and physics).

A more precise description of tasks and roles was waived as it is not possible to change legal regulations (public services law, etc.) with any degree of expediency. The question as to the roles and structures these persons will adopt on completion of the course therefore remains open. This bears risks (role ambiguity) but also chances (to engage in definition and design, especially during and by means of the course).

The following task description for subject-related education managers was defined as a starting point to be specified in greater detail at a later date (cp. Krainer, Fischer & Wallner, 2006):

> Subject co-ordinators, members of regional networks, etc., fulfil at a regional level important tasks relating to the discussion of fundamental questions of education and its understanding, measures pertaining to cross-school quality development and assurance, the co-ordination of professional exchange, the organisation of courses relating to subject and subject didactics, the initiation of regional development work and the mutual co-ordination with regard to the requirements of different schools and school types.

They should initiate and support processes in their respective federal states with the aim of critically appraising subject related and interdisciplinary teaching methods for each subject, and to (further) develop new methods. At the same time it is important to visualise and propagate innovations and "good practice," as well as informing and discussing new scientific findings, current issues and challenges such as educational standards (Krainer, Kühnelt, Peschek & Wintersteiner, in press). This will enable an expedient and more efficient response to the challenges and issues arising in the education system both from schools ("bottom-up") and education authorities ("top down"). Subject-related education managers therefore gain special significance with regard to "interface management." They are an important prerequisite for enabling a location-related and super-regional standard education in the subject groups and in regional subject groups and networks.

The great competence variety of subject-related education managers and a lack of empirical experience prior to the start of the course made it impossible to develop a more detailed specification of the competence model (e.g., as developed for mathematics teachers over the past years within the COACTIV framework, cp. Brunner et al., 2006). However, first steps in this direction are feasible within the framework provided by course implementation, evaluation and accompanying research. Five preliminary competence areas—"In-service Education," "Consulting," "Co-operation and Networking," "Subject-political Issues," and "Analysis and Reflection"—have been constructed by the accompanying research team in co-operation with the course team. First experience with this concept development are reported in Chapter 4.

COURSE CONCEPT AND CONTENT

The main themes of the course (cp. Fischer & Krainer, 2006) are—in accordance with its name—*Education, Management* und *Subject (didactics)*. The course follows von Hentig (2004) in as much as *education* is not only understood as self-education but as a conscious responsibility for and active participation in the shaping of society. Education aims at improving the ability to act and at assuming responsibility as well as for the individual and the collective. *Management* means to make or induce decisions and communicate this clearly. As a perpetual process concerning individuals and the collective, education needs ongoing decisions and actions as much as it needs their critical assessment. In this context, teachers (at schools, colleges, universities, etc.) are the most important education managers appointed by society, and the goal is to strengthen their collective actions as a community and profession. Organising a profession requires a minimum structure. A lot speaks in favour of using *subjects* as a structuring element in accordance with their development in schools, universities, etc.. The subjects represent points where knowledge, competence, beliefs and culture crystallise (cp. Neubrand, 2006; Leder, Pehkonen & Törner, 2003). The scientific disciplines concerned with teaching and learning a subject are the *subject didactics.*

The *modules* of the course (40 credits in the European Credit Transfer System) follow the three main themes and focus on the following issues and subjects:

- *Subject didactics module:* subject didactics and pedagogy in a subject-related and interdisciplinary context under consideration of current issues and challenges (e. g. results from various studies, education standards).
- *Interface management module:* Self-image and design of one's own role as subject-related education manager, project management and networking.
- *Education module:* Education and teaching with regard to current and future societal demands and relevance.

Compulsory optional subjects supplement and enhance the modules. Participants are required to write two *reflective papers* on a practice-related case, one related to the corresponding "subject didactics" and one related to the "interface management" module. This could be, for instance, the organisation and evaluation of a teacher education course or the co-ordination of an education standard project in a district, but generally not a project for improving one's own teaching skills. The primary goal of the course is not that teachers become better teachers, but rather the new and further development of their competences in regional education management.

THE CONCEPT OF ACCOMPANYING RESEARCH

The research accompanying the "Subject-related Education Management" course is based on a longitudinal design that analyses the course with regard to input, process flow and output (cp. Mayr & Müller, 2006). The theoretical background is based on a *supply-use model* (Helmke & Weinert, 1997) that can be explained in simplified terms as follows: Participants begin the course with specific *initial preconditions* (expectations, experience, interests, skills,...) and find various predefined *learning opportunities* (in the form of learning materials, persons, information,...). They *use* these depending, on the one hand, on their own preconditions and, on the other hand, on the quality of the learning opportunity. The individual learning benefit gained from each course (in the sense of broadening knowledge bases, changing convictions,...) is considered as dependent on initial preconditions, learning opportunities and the use made thereof (cp. Mayr & Müller, 2006; see also Figure 15.1).

The competence acquired through the course (a broad meaning of the term competence is assumed; cp. Allemann-Ghionda, 2006) should be reflected in professional skills and actions and therefore represent a contribution to the further development of each individual's professional skills and, as a consequence, to teaching as a profession and to the education system.

A *longitudinal study* containing semi-standardised "knowledge tests" in addition to questionnaires and interviews (in order to record participants' self-disclosure data) will be carried out in parallel. It is designed partially on established theoretical concepts and the corresponding methods of collecting data, and also includes new, target group-related developments constructed, for example, on the basis of interviews with the course management team.

The study investigates all participants' development in the course. Figure 15.1 shows the study plan and the main areas of investigation for the two-year university course.

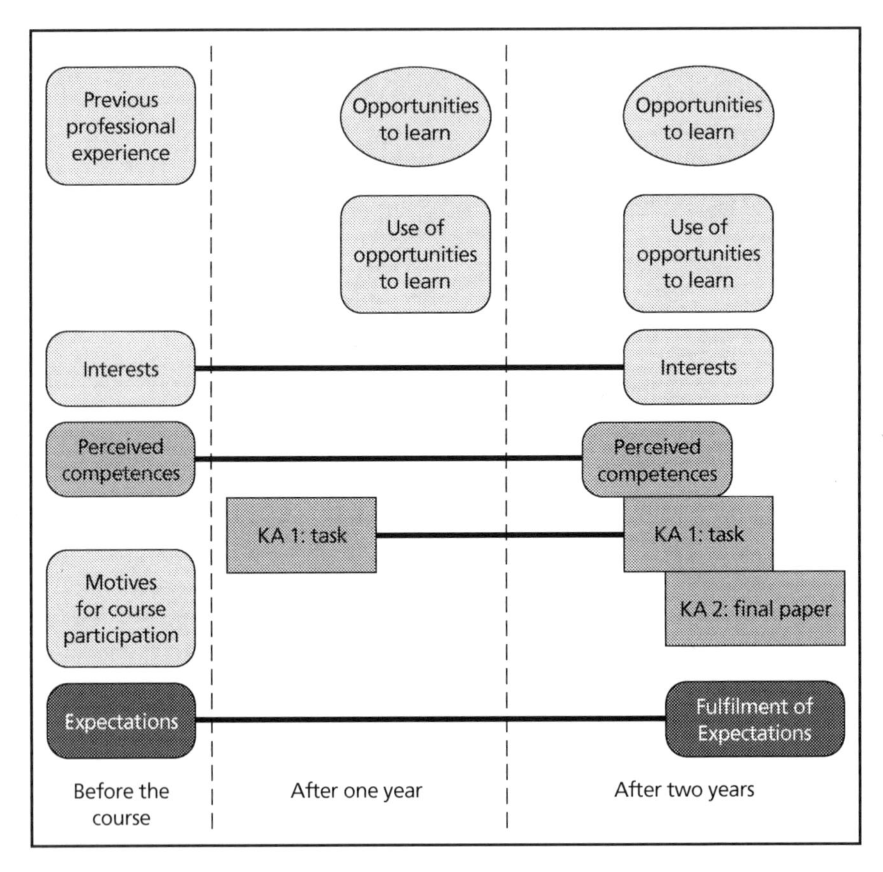

Figure 15.1 Study design for the course with data collection time points and selected variables (cp. Mayr & Müller, 2006).

Important learning prerequisites of participants were recorded *before the course commenced* and are detailed briefly in the following:

- *Previous professional experience*: in addition to their immediate teaching experience, participants were asked about experience that could be of particular relevance to subject-related educational management. This could be, for instance, experience in instructional and school development, in the evaluation of learning environments, the evaluation of educational measures or in organising and implementing in-service education courses for teachers.
- *Motives for course participation:* a questionnaire on participation motives was designed for this purpose that is oriented on the theory of self-determination by Deci und Ryan (1992) and assumes that

the choice of educational course is determined by multiple motive categories.

- *Expectations from the course*: these refer to expectations with regard to expanding competences and in particular methodological and didactical preferences.
- *Interests*: These were captured on the one hand by means of an abridged version of a "General Interest Structure Test" by Bergmann and Eder (1992) and a questionnaire containing possible fields of activity of education managers, such as in-service education, consulting, co-operation and networking, subject-political activities, and reflection and analysis, etc. Furthermore, participants were asked not only to estimate their interest in these fields but also their *competence* (cp. also Table 15.1).
- *Personality*: an abridged version of a "Big-Five" personality inventory was used (Rammstedt et al., 2005). Big Five personality markers should be seen as "basic tendencies" for attitudes, convictions, actions, etc. (cp. McCrea et al., 2000). This personality concept has proved suitable for the moderate prognosis of learning processes and results (De Raad & Schouwenburg, 1996) or, in particular, the prognosis of "successfully" coping with the teaching profession (cp. Mayr & Neuweg, 2006).

Immediately after the course started, participants were assigned the task of writing a newspaper article on the significance of the subject for which they would be responsible as education managers (see Fig. 15.1, KA 1: Task) and to document their approach to solving the task. Articles are subsequently evaluated by external persons with a view to various content-related and formal quality aspects. Data collections of this nature are repeated at subsequent stages to obtain indicators for change. A selection of projects executed by participants is also analysed (see Fig. 15.1, KA 2: Final paper). The intention is to obtain third-party appraisals—in addition to the self-perception—for those participants with a tendency to provide socially desirable responses.

Statements are collected at the end of the second and fourth semester with regard to the opportunities to learning provided by the course and how they were used. Guideline-based interviews with course team members aimed at viewing relevant aspects of learning opportunities are conducted for the purpose. These aspects are intended to form a basis for constructing a questionnaire to capture learning opportunities and the use made thereof.

Occupation-related interests and competences recorded at the start are again queried or covered by tasks at the end of the course. The intention is to indicate changes that could be interpreted as a result of the course. Participants are also asked to indicate how far their expectations of the course were fulfilled.

A follow-up study aimed at establishing the long-term effects of the course is planned once the course has been completed. Amongst other things, the follow-up study will enable to find out to which extent participants are able to make use of the qualifications gained from the course in their professional careers.

INITIAL EMPIRICAL FINDINGS FROM ACCOMPANYING RESEARCH

In the following section, exemplary results from the survey held *before* the course are presented and interpreted on the basis of the course concept. It is also intended to show how findings from the accompanying research—in the sense of a formative evaluation—can provide practical indicators for the design of the current course.

The questionnaires on participations motives, interests and competences are currently in the trial phase and will need to be checked for reliability and validity in the course of further studies.

Sample

A total of 144 prospective subject-related education managers were surveyed by means of an online questionnaire before the course commenced. 140 evaluable datasets were returned. The majority of participating teachers teach at main schools (60%), whilst 40% teach at schools for higher general education ("Gymnasium"). Around 40% of the 140 course participants are female, and the same number have already held or organised in-service education courses for teachers. The participants are distributed more or less equally among the subject modules German (35%), mathematics (30%) and natural sciences (35%). This heterogeneity among participants (school type, subjects, previous experience,...) in combination with the size of the course represents both a challenge and a learning opportunity.

Results: Motives for Choosing the Course

The majority of participants state intrinsic motivation for course participation as opposed to extrinsic motivation. This indicates good preconditions for learning—at least from a motivational theory viewpoint. Figure 15.2 shows some examples of motives for choosing the course. These items are based on a five-level rating scale (1 = does not apply, 5 = applies in full).

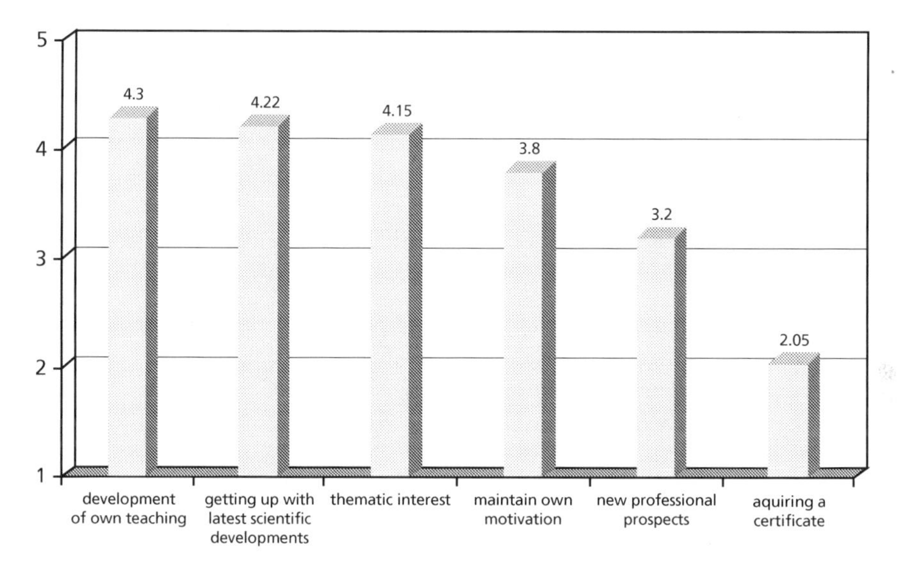

Figure 15.2 Selected motives for choosing the course "Subject-related Education Management."

The most significant motive for course participation is to further develop one's own teaching skills (M: 4.3, SD: 0.84). However, this is not a primary intention of the course and could therefore lead to inappropriate motivation. On the other hand, only 11% of participants stated this motive as the only main motive for choosing the course. This group runs the risk of disappointed expectations in course content if their perceptions of the course and the course itself are incompatible. All other participants display a variety of motives that are compatible with the goals of the course. This is also reflected by the areas of interest assessment in Table 15.1. Tendencies towards the motives of "getting up-to-date with the latest scientific developments" (M: 4.22, SD: 0.91) and "thematic interest in the course" (course content and concept) (M: 4.15, SD: 0.81) are high at the start of the course. The course was also chosen to "maintain motivation in my profession as a teacher" (M: 3.80, SD: 1.22) or to acquire "new professional prospects" (M: 3.20, SD: 1.15) (in addition to teaching). Extrinsic motives, such as acquiring a certificate (M: 2.05, SD: 1.16) are less pronounced overall.

Result: Interests and Competences

Table 15.1 shows the participants' assessment of interests and competences for the areas "In-service Teacher Education," "Consulting," "Co-

TABLE 15.1 Self-Assessed Interests and Competences

Dimensions and item examples	Interest mean (SD)	Competence mean (SD)
In-service Teacher Education: 7 items; α (interest): 0.76/ α (competence): 0.80	3.85 (0.63)	3.36 (0.75)
Organising in-service education for teachers	4.12 (0.95)	3.75 (1.20)
Conducting in-service education for teachers	4.08 (1.00)	3.47 (1.18)
Designing a subject-related in-service education concept for teachers	3.84 (1.10)	2.90 (1.12)
Evaluating courses of in-service education for teachers	3.27 (1.11)	2.88 (1.17)
Consulting: 5 items; α (interest): 0.80/α (competence): 0.82	4.09 (0.66)	3.57 (0.78)
Supporting teachers in teaching-related problems	4.21 (0.89)	3.79 (1.03)
Coaching teachers in subject-related and didactical issues	4.05 (0.91)	3.58 (0.96)
Implementing subject-related perspectives to school development processes	4.02 (1.00)	3.44 (1.14)
Co-operation and Networking: 7 items; α (interest): 0.80/ α (competence): 0.82	3.84 (0.66)	3.19 (0.77)
Collaborating with teacher education institutes	4.20 (0.83)	3.55 (1.02)
Maintaining contacts to scientific experts	4.17 (0.89)	3.17 (1.17)
Initiating co-operation between schools and educational administration	3.59 (1.12)	2.99 (1.22)
Subject political issues: 5 items; α (interest): 0.78/ α (competence): 0.82	3.92 (0.72)	3.33 (0.83)
Increasing school awareness for the importance of the subject	4.06 (0.99)	3.65 (1.05)
Pointing out the importance of a subject to those who are not familiar with this subject	3.98 (0.89)	3.52 (1.04)
Making publicity for the subject	3.75 (1.05)	3.16 (1.12)
Analysis and Reflection: 5 items; α (interest): 0.76/ α (competence): 0.82	3.73 (0.71)	3.14 (0.85)
Identifying weak points in the education system	3.96 (0.96)	3.36 (1.09)
Preparing scientific knowledge for teachers	3.80 (0.98)	3.18 (1.07)
Deducing practical consequences from international school assessments	3.52 (1.04)	2.91 (1.06)

SD = standard deviation; α (Alpha) = reliability (Cronbach).
Scale interest: "I like to do the following activities": 1 = not at all; 5 = very much
Scale competence: "I feel competent in the following activities": 1 = not at all; 5 = very much

operation and Networking," "Subject-political Issues," and "Analysis and Reflection" mentioned previously.

The exemplary items are intended to indicate how the scales are composed. The degree of the reliability coefficients can be assessed as satisfactory (Cronbachs Alpha between 0.76 and 0.82). However, it should be noted that the scales should not be viewed as selective dimensions. Intercorrelations between the scales fluctuate between 0.48 and 0.76 for the interests scale and between 0.63 and 0.79 for competences. The scales for "In-service Teacher Education" and "Co-operation and Networking" show the greatest correlation between interests and competences. Future adjustments to items and conceptional considerations relating to the areas of competence of subject-related management will be required in particular for these two scales.

It can be seen on the five-level scale that assessments of the *areas of interests* lie between 3.73 (SD: 0.71) for "Analysis and Reflection" and 4.09 (SD: 0.66) in the area of "Consulting." Thus, all mean values are higher than the scale value three of a five-stepped additive scale, and not one item value is below this mark. Interest in the area of "In-service Teacher Education" lies in particular in "organising" (M: 4.12, SD: 0.95) and "conducting in-service teacher education" (M: 4.08, SD: 1.00). Investigating the quality and efficacy of in-service education measures ("evaluating in-service teacher education") was of little interest to participants—they declared it to hold the least interest value (3.27, SD: 1.11). In contrast, "Consulting" holds a high level of interest, in particular supporting colleagues in "problems in practical teaching," whereby this item scores the highest interest value (M: 4.21, SD: 0.89).

Interest in the area of "Co-operation and Networking" is equally as high, in particular in conjunction with co-operating with teacher education institutes (M: 4.20, SD: 0.83) and scientific experts (M: 4.17, SD: 0.89). "Increasing school awareness for the importance of the subject" gained the most interest in the field of "Subject-political Issues" (M: 4.06, SD: 0.99).

The identification of weak points in the education system achieved the highest score in the field of "Analysis and Reflection" (M: 3.73, SD: 0.71). It is assumed that "evaluating courses of in-service teacher education" had the lowest score because the output-related evaluation of one's own activities or those of colleagues is considered unpopular. Input-related and process-related activities such as support, education, organisation or co-operation, on the other hand, are considered as much more interesting.

The rating sequence of mean values for the areas of *competence* is virtually the same as for interests—the areas of "Analysis and Reflection" (M: 3.14, SD: 0.85) and "Consulting" (M: 3.57, SD: 0.78) score the lowest and highest values respectively. In addition to "Analysis and Reflection," competence self-assessment in the field of "Co-operation and Networking" is also relatively low (M: 3.19, SD: 0.77). "Supporting teachers in teaching-related problems" scores the highest value for competences (M: 3.79, SD: 1.03). This item, together with "organising in-service education for teachers," "evaluating in-service education," and "Increasing school awareness for the importance of the subject" shows the least difference between interest and competence values. At the same time, these items (with the exception of "evaluating in-service education") scored the highest competence values (3.65 to 3.79).

The greatest differences between interest and competence scores are seen in conjunction with the items "designing a subject-related in-service education concept for teachers" and "maintaining contacts to scientific experts." This is where the course would seem to hold the greatest potential for development—based on participants' self-assessment. This applies in particular to the item "maintaining contacts to scientific experts," as it scores one the highest interest values and the course is aimed at achieving a close co-operation between representatives from practice and science. Further favourable preconditions for developments in the course are particularly apparent where participant interest is high—as in the field of "Consulting."

The results allow the formulation of a hypothesis that tendencies have experienced a positive longitudinal change over the two years, particularly in low-scoring areas of competence (such as designing in-service education concepts, evaluating in-service education, quality assurance and practice-oriented interpretation of scientific studies), as they concern the intended content of the course (whereby tendencies differ from subject to subject).

The degree of assessment in the areas of interest and competence was significantly influenced by the variable *participants' own experience in organising and implementing in-service education for teachers*. Table 15.2 shows that those teachers with experience in in-service teacher education (41% of participants) assess their competence in particular as higher, in addition to their interest. There is no significant difference in the interest shown for the areas "Subject-political Issues" and "Analysis and Reflection".

It is by all means plausible that participants with experience in in-service teacher education should assess their competence in the fields of "In-service Education," "Consulting," and "Co-operation and Networking" as higher. However, it is interesting from a longitudinal viewpoint to see how the self-assessed competences change in relation to participants' interests and previous experience. The previously mentioned analysis of the "newspaper article task"—as an aspect of third-party competence assessment—should provide further insight during the investigation into changes. The task is presented to the participants at the end of the course for post-editing.

Gender-specific differences relating to self-assessment are only found for the scale "Co-operation and Networking": women consider themselves slightly more competent than men (female: $M = 3.4$, $SD = 0.74$; male: 3.0, SD: 0.74; $t(137) = 2.82$, $p = 0.005$). No significant group differences are found between subjects, school types and other relevant variables.

Attention is drawn to the area of conflict between course concept, course implementation and the basic conditions that were already of relevance

TABLE 15.2 Mean Value Differences with Regard to Reported Experience in Education Measures for Teachers

	Experienced		Non-Experienced		t-Test		
Interest	M	SD	M	SD	df	t	p
In-service Education	3.99	0.55	3.75	0.66	138	2.21	0.03
Consulting	4.25	0.67	3.99	0.64	138	2.29	0.02
Cooperation and Networking	4.05	0.54	3.71	0.69	138	3.02	0.00
Subject Political Issues	3.99	0.76	3.87	0.68	138	0.96	0.33
Analysis und Reflection	3.86	0.73	3.64	0.67	138	1.83	0.07
Competence							
In-service Education	3.62	0.66	3.17	0.76	138	3.64	0.00
Consulting	3.85	0.72	3.38	0.77	138	3.65	0.00
Cooperation and Networking	3.47	0.68	2.95	0.75	138	4.19	0.00
Subject Political Issues	3.59	0.74	3.16	0.83	138	3.14	0.00
Analysis and Reflection	3.42	0.76	2.93	0.86	138	3.49	0.00

Scale interest: "I like to do the following activities": 1 = not at all; 5 = very much
Scale competence: "I feel competent in the following activities": 1 = not at all; 5 = very much

at the start of the course: the particular challenge facing the course team should be seen in the fact that the majority of participants defined the occupational profile and professional prospects for subject-related education managers as unclear at the start of the course. This should be no surprise, given that significant basic conditions for subject-related education management in Austria are yet undefined.

The assessments selected exemplarily in Table 15.3 make it obvious that participants' uncertainty with regard to unclear job profiles and prospects for subject-related education managers is reflected in the initial assessment of the course content.

TABLE 15.3 Assessment of Course Goals and the Tasks Facing Subject-Related Education Managers

Items	Percentage of participants who partly or fully agree
I am not clear about the aims of the course	53%
For me it's unclear which job profiles and prospects for subject-related education managers are relevant	78%

Scale: 1 = fully agree; 5 = do not agree

SUMMARY AND OUTLOOK

The *course "Subject-related Education Management"* qualifies teachers to adopt tasks relating to subject-related middle management in the nine federal states of Austria. Evaluation and accompanying research are all the more important as the concrete roles to be played by these persons are yet still open.

Further development of one's own teaching methodology is the most significant motive for participants to take part in the course. Appropriate processes for reflection and negotiation during the course should be organised as this is not a primary course goal. However, nearly all participants (just under 90%) stated at least an additional major motive for their participation. It is also conducive that intrinsic aspects of motivation make up the vast majority. The course will offer the majority of participants (just under 60%) a first opportunity to organise and implement an in-service education measure (or similar practical activity) for teachers themselves (and to document and evaluate it). Appropriate support and exchange—including among teachers—will be of importance.

Five areas of competence—"In-service Education," "Consulting," "Co-operation and Networking," "Subject-political Issues," and "Analysis and Reflection"—were constructed as an initial approach to defining a concept for education manager competence. Participants were asked to assess their corresponding interests and competences before the course commenced. Interests scored consistently higher marks than competences. The fields of "Consulting" and "Analysis and Reflection" score the highest and lowest values respectively with regard to both interest and competence. The greatest differences between interest and competence scores are seen in conjunction with the items "Designing a subject-related in-service education concept for teachers" and "Maintaining contacts to scientific experts." This is where the course would seem to hold the greatest potential for development—based on the self-assessment of course participants.

Initial findings from accompanying research show that there are no significant differences between teachers with regard to the self-assessment of interests and competences in relation to subject area, gender or school type. However, differences exist in the individual experience in organizing and conducting in-service teacher education. Participants with experience of this nature assess their interest and, in particular, their competence as higher. This stresses the significance of individual support and exchange of experience among participants, in particular in conjunction with colleagues with a wealth of previous experience.

The relatively large amount of ambiguity regarding the aims of the course and the extreme vagueness of participants' future tasks as subject-related education managers is hardly surprising, given the lack of definition of functions and roles. However, it is important to discuss these issues during the course and to convince the participants and other relevant parties how urgent the need for regulations is. Therefore, the course team, the participants and, in particular, the education authorities will need to display adequate "education management." In this regard it is the entire system that is learning, not just the course participants.

NOTES

1. This paper is a slightly modified version of Krainer & Müller (2007): see references.
2. IMST (Innovations in Mathematics, Science and Technology Teaching) is an Austria-wide initiative for improving the teaching of mathematics, science and technology (see also: http://imst.uni-klu.ac.at).

REFERENCES

Allemann-Ghionda, C. & Terhart, E. (Hrsg.) (2006). Kompetenzen und Kompetenzentwicklung von Lehrerinnen und Lehrern: Ausbildung und Beruf. *51. Beiheft der Zeitschrift für Pädagogik.* Weinheim: Beltz.

Bergmann, C., Eder, F. (1992). *Allgemeiner Interessen-Struktur-Test (AIST). Umwelt-Struktur-Test (UST).* Testmanual. Weinheim: Beltz.

Brunner, M., Kunter, M., Krauss, S., Klusmann, U., Baumert, J., Blum, W., Neubrand, M., Dubberke, T., Jordan, A., Löwen, K. & Tsai, Y.-M. (2006). In M. Prenzel & L. Allolio-Näcke (Eds.), *Untersuchungen zur Bildungsqualität von Schule. Abschlussbericht des DFG-Schwerpunktprogramms* (pp. 54–82). München: Waxmann.

Deci, E. L. & Ryan, R. M. (1993). Die Selbstbestimmungstheorie der Motivation und ihre Bedeutung für die Pädagogik. *Zeitschrift für Pädagogik 39,* 223–228.

De Raad, B. & Schouwenburg, H. C. (1996). Personality in learning and education: A review. *European Journal of Personality* 10, 303–336.

Fischer, R. & Krainer, K. (2006). Bildung—Management—Fach. Utopie und Realität. In Universitätslehrgang "Fachbezogenes Bildungsmanagement" (Ed.), *Lehrgangsmappe, Einführung in den Lehrgang und Modul Bildung* (pp. 1–10). Klagenfurt: IUS.

Helmke, A. & Weinert, F. E. (1997). Bedingungsfaktoren schulischer Leistungen. In F. E. Weinert (Ed.), *Psychologie des Unterrichts und der Schule,* Vol. 3, (pp. 71–176). Göttingen: Hogrefe.

von Hentig, H. (2004[5]). *Bildung. Ein Essay.* Weinheim: Beltz.

Krainer, K. (2007). Die Programme IMST und SINUS: Reflexionen über Ansatz, Wirkungen und Weiterentwicklungen. In D. Höttecke (Ed.), *Naturwissenschaftliche Bildung im internationalen Vergleich* (pp. 20–48). Münster: LIT.

Krainer, K., Fischer, R. & Wallner, B. (2006). Universitätslehrgang "Fachbezogenes Bildungsmanagement." *IMST3 Newsletter 5*(16), 15–16. http://imst.uni-klu.ac.at/materialien/2006/712_Newsletter_16.pdf (06.03.2007).

Krainer, K., Kühnelt, H., Peschek, W. & Wintersteiner, W. (in press). Fachbezogenes Bildungsmanagement und Standards. In P. Labudde (Ed.), *Bildungsstandards am Gymnasium: Korsett oder Katalysator?* Bern: h.e.p. Verlag.

Krainer, K. & Müller, F. H. (2007). Fachbezogenes Bildungsmanagement: Konzeption eines Lehrgangs und erste Befunde der Begleitforschung. In A. Peter-Koop, & A. Bikner-Ahsbahs (Eds.), *Mathematische Bildung—Mathematische Leistung* (pp. 79–95). Hildesheim: Franzbecker.

Leder, G., Pehkonen, E. & Törner, G. (Eds.) (2003). *Beliefs: A hidden variable in mathematics education?* Dordrecht: Kluwer.

McCrae, R. R., Costa, P. T., Ostendorf, F., Angleitner, A., Hrebíková, M., Avia, M. D., Sànz, J., Sanchez-Bernados, M. L., Kusdil, M. E., Woodfield, R., Saunders, P. R. & Smith, P.B. (2000). Nature over nurture: temperament, personality, and life span development. *Journal of Personality and Social Psychology* 78 (1), 173–186.

Mayr, J. & Neuweg, G. H. (2006). Der Persönlichkeitsansatz in der Lehrer/innen/forschung. To be published in U. Greiner & M. Heinrich (Eds.), *Schauen, was 'rauskommt. Kompetenzförderung, Evaluation und Systemsteuerung im Bildungswesen* (p. 183-206). Münster: LIT.

Mayr, J. & Müller, F.H. (2006): Begleitforschung zum Universitätslehrgang "Fachbezogenes Bildungsmanagement" (fBM). http://fbm.uni-klu.ac.at/begleitforschung/ (25.04.2007).

Müller, F. H. & Mayr, J. (2007). Was macht Lehrer/innenfortbildung wirksam? Begleitforschung zu den Universitätslehrgängen PFL, ProFiL und fBM. In K. Krainer, J. Mayr, R. Müller & A. Turner (Eds.), *Jahresbericht 2006 des Instituts für Unterrichts- und Schulentwicklung* (pp. 57–60). Klagenfurt: IUS. http://ius.uni-klu.ac.at/publikationen/dateien/jahresbericht_2006.pdf (15.05.2007).

Neubrand, M. (2006). Professionelles Wissen von Mathematik-Lehrerinnen und Lehrern: Konzepte und Ergebnisse aus der PISA- und der COACTIV-Studie und Konsequenzen für die Lehrerausbildung. In F. Kostrzewa (Ed.), *Lehrerbildung im Diskurs* (pp. 53–72). Eitorf: gata-Verlag.

Rammstedt, B., Koch, K., Ingwer, B. & Reitz, T. (2005). Entwicklung und Validierung einer Kurzskala für die Messung der Big Five Persönlichkeitsdimensionen in Umfragen. *ZUMA Nachrichten*, Nr. 5. http://www.gesis.org/ZUMA (15.06.2006).

CHAPTER 16

INTEGRATING INTUITION

The Role of Concept Image and Concept Definition for Students' Learning of Integral Calculus

Bettina Rösken
University of Duisburg
Germany

Katrin Rolka
University of Dortmund
Germany

Abstract: In this paper we analyze students' conceptual learning regarding the notion of the definite integral. By means of a comprehensive questionnaire, students' concept image and concept definition, as mentioned in the model by Tall and Vinner (1981), were ascertained as well as the corresponding problem solving competence. Participants in this study were 24 students in grade 12 of a German secondary school. The results indicate that definitions play a marginal role in students' learning whereby intuition inherent in concept images dominates the conceptual learning. Based on these subjective convictions, intended and realised knowledge may deviate from each other and might cause difficulties for students.

We use this pre-introduction to establish a relationship between the content of this paper and our supervisor, colleague and friend to whom is dedicated this volume. In his research, Günter Törner has always been guided by his dual viewpoint, on the one hand, as mathematician and, on the oth-

The Montana Mathematics Enthusiast, pages 231–259

er hand, as mathematics educator. He has always connected his profound mathematical background to didactical approaches in a very fruitful and meaningful way. In particular, he has always rejected pure rote learning and instead emphasized the deep understanding of mathematical notions in order to make sense and grasp the meaning of mathematics beyond definitions. This also includes the development of own ideas and rich images. The focus of our paper follows this approach by analyzing students' learning of the definite integral using Tall and Vinner's (1981) model of concept image and concept definition.

INTRODUCTION

Already in the last century, the famous mathematician Poincaré (1908/52) described impressively the contrast between the nature of mathematics, on the one hand, and students' difficulties to grasp mathematical ideas, on the other hand, when asking the following:

> How is it that there are so many minds that are incapable of understanding mathematics? Is there not something paradoxical in this? Here is a science which appeals only to the fundamental principles of logic [...], and yet there are people who find it obscure, and actually they are the majority. (p. 117)

In order to explain this phenomenon, students' learning has been studied from different perspectives, focusing especially on subjective conceptions related to the objective mathematical ones. Bauer (1995), for example, elaborates on individual and subjective aspects of students' learning of mathematics and poses the general question, *What is the relation between mathematics and human thinking?* Addressing also findings from cognitive psychology, this more or less philosophical question can be refined as follows, *What possibilities does our cognitive architecture offer, what restraints does it entail?* More concretely, *Which habits do hinder an adequate formation of mathematical concepts?* Everyday concepts cannot always be uniquely characterized; consequently, limits of concepts are fuzzy. In contrast, mathematical concepts are ideal ones, explicitly concretized in formal definitions. As an example for this decisive but not always transparent distinction serves Figure 16.1.

The immediate categorization of both figures as triangles describes an essential capacity of our brain, simultaneously constituting difficulties on the formal mathematical level. Hence, mathematical concept formation is different from human thought processes, mostly guided by heuristic strategies (Tversky & Kahneman, 1974). Compared to everyday life, the role of definitions in mathematics is critically raised by Freudenthal's (1973) say-

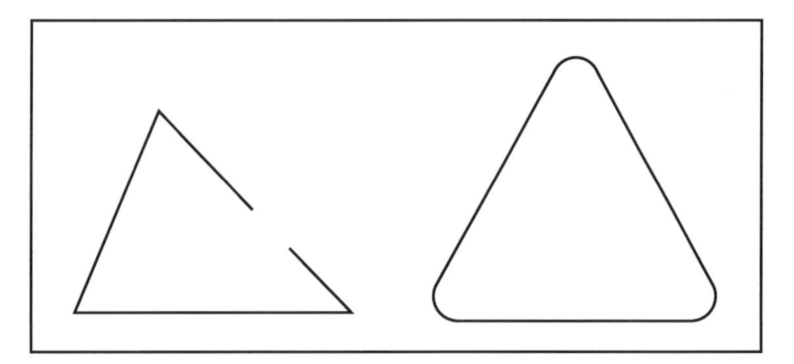

Figure 16.1 Examples for triangles in everyday contexts but not in mathematical ones (Hartland, 1995, p. 217).

ing: "Nobody defines what it means to breathe, to walk, to fall, to swim, and nevertheless people learn to do so [...]" (p. 317). In the work of mathematicians, however, definitions play an essential role.

The interplay between subjective conceptualizations and formal definitions of mathematical concepts is discussed in this paper by employing Tall and Vinner's (1981) model of concept image and concept definition. We will further elaborate on these dual constructs in the next chapter. Particularly, we will point out how this model can be applied to analyze students' conceptions of a specific mathematical notion, that is, the definite integral.

STUDENTS' LEARNING OF MATHEMATICAL CONCEPTS

Students' ways of mathematical thinking, particularly cognitive aspects, have been analyzed from different perspectives and with different emphasis. Nevertheless, these approaches, whether they are more cognitive based (Harel & Sowder, 2005) or more visual oriented (Even & Tirosh, 2002; Presmeg, 2006), share a constructivist view on students' learning and especially refer to students' intuitive thinking as decisive part of acquiring new knowledge. Fischbein (1987) examined students' individual concept formation and elaborated on how intuitive models influence the learning of mathematics:

> It is very well known that concepts and formal statements are very often associated, in a person's mind, with some particular instances. What is usually neglected is the fact that such particular instances may become, for that person, universal representatives of the respective concepts and statements and then acquire the heuristic attributes of models. (p. 149/150)

During their learning of mathematics, students are faced with a wide range of information. How they choose to integrate new mathematical aspects and develop concepts will also depend on their beliefs, values and previous experiences. Consequently, the field of students' subjective concept formation can also be analyzed from the viewpoint of epistemological beliefs (Köller, Baumert & Neubrand, 2000). In this regard, we refer to Schoenfeld's (1998) notion of beliefs as "mental constructs that represent the codifications of people's experiences and understandings" (p. 19). That is, acquiring a mathematical concept is mainly influenced by ontological and epistemological assumptions that have become manifest through students' hitherto learning history.

In particular, Tall and Vinner (1981) have dealt with creative processes in students' learning of mathematics. Their model of concept image and concept definition allows for analyzing students' representations of mathematical concepts. Vinner formulated the terms concept image and concept definition in 1980. Initially, he analyzed how students conceive simple geometric figures and the relations between them (Vinner & Hershkowitz, 1980). The aforementioned terms were introduced in order to categorize students' difficulties emerging because of their tendency to favor prototypical learning. Students focused on typical examples and neglected information given by the mathematical definition. Geometrical figures like a rectangle are represented as typical examples and students therefore have difficulties to consider, for example, a square as a rectangle. The observations made during that study are supported by comparable findings of psychologists, particularly Rosch's (1975) research on prototypes. Contrary to mathematics, concepts in everyday life are learned through examples rather than abstract rules.

At that time, Tall investigated students' cognitive conflicts when learning calculus, in particular when dealing with limits and continuity. The findings of both researchers, Tall and Vinner, led to a common paper in 1981 concerned with concept image and concept definition on which we will elaborate more deeply in the following. The approach presented there has been refined by both authors, partly in separate papers, and also by other researchers within the community and remarkably, the notion is still relevant in the current research (Ouvrier-Buffet, 2006). Another study published at the same time and worth mentioning here, is the one by Cornu (1981) who labels the subjective representations of students as *modèles propres*. His objective has also been to show how these individual concept formations hinder adequate learning of the intended mathematical theory.

Concept Image and Concept Definition

Tall and Vinner (1981) contrast the cognitive processes of conceptual learning with structural features and particularities of mathematics:

> The human brain is not a purely logical entity. The complex manner in which it functions is often at variance with the logic of mathematics. It is not always pure logic that gives us insight, nor is it chance that causes us to make mistakes. To understand how these processes occur, both successfully and erroneously, we must formulate a distinction between the mathematical concepts as formally defined and the cognitive processes by which they are conceived. (p. 151)

By using the constructs of concept image and concept definition, Tall and Vinner differentiate aspects of mathematical knowledge that, on the one hand, are given by formal definitions and, on the other hand, by subjective constructions. Figure 16.2 illustrates these two constructs and the ideas around them.

Tall and Vinner (1981) use the term concept image "to describe the total cognitive structure that is associated with the concept, which includes all the mental pictures and associated properties and processes" (p. 152).

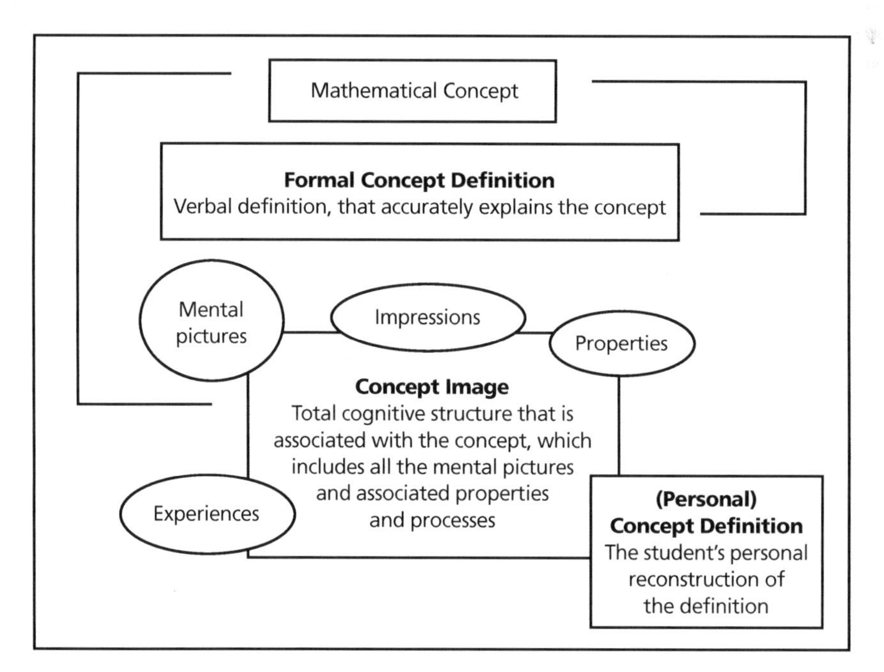

Figure 16.2 Exemplification of concept image and concept definition

The concept image may also include a (personal) concept definition as an individual reconstruction of the mathematical one. Further, Tall and Vinner use the term evoked concept image to describe the currently activated part of the concept image. Interestingly, the concept image is not necessarily consistent and coherent regarding its content; it might also include contradictory aspects that students are not aware of.

Finally, we would like to emphasize the importance of this model for learning and teaching mathematics. Explanations for concepts will easily be forgotten if students are not able to develop own ideas and associations. Learning a new concept requires forming a comprehensive concept image but one should keep in mind that maybe important aspects of a mathematical concept are not adequately represented. In the following section we further elaborate on this idea.

Students' Alternative Conceptions

In order to show the significance of the model, Vinner (1994) contrasts the concept formation in technical contexts with everyday life contexts. Most everyday concepts are acquired as ostensive definitions and, to that effect, formal definitions are of inferior relevance. In everyday contexts, concept images play a crucial role whereas in technical contexts definitions are often essential. Specific to mathematics, it is mostly indispensable to consider all aspects of a definition. For example, if students are asked to examine a continuous (real-valued) function on a closed interval regarding relative extrema they tend to neglect the ones in the endpoints of the interval. In that case students' thinking is dominated by the concept image of horizontal tangent and they do not take into account that the corresponding condition $f'(x) = 0$ is only relevant for the open interval.

By the given model, Vinner (1994) explains why students' misconceptions, like the one mentioned before, occur in learning situations. He shows that during the process of concept formation the relation between the concept image and concept definition is reciprocal (Fig. 16.3):

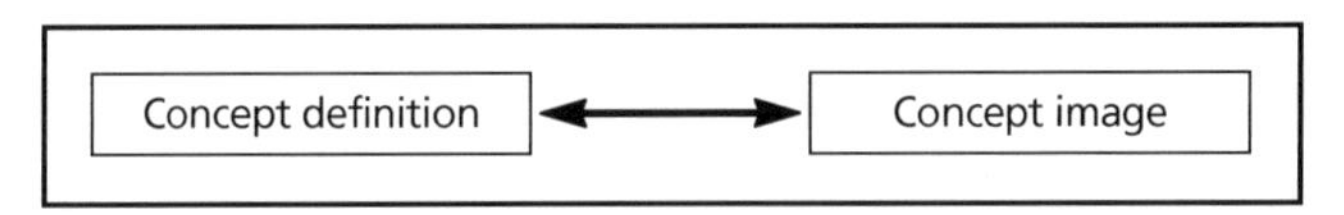

Figure 16.3 Relationship between concept image and concept definition during concept formation (Vinner, 1994, p. 70).

In contrast to that situation, teachers assume a one-way relationship from the concept definition to the concept image, like it is shown in Figure 16.4:

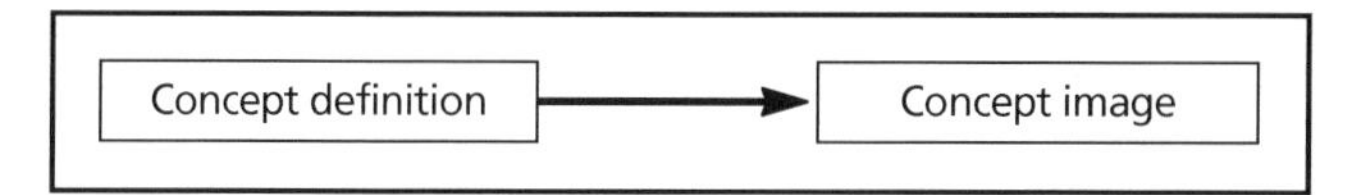

Figure 16.4 Expected relationship between concept image and concept definition during concept formation by teachers (Vinner, 1994, p.70).

The assumption that the concept definition controls the content of the concept image would indicate an adequate concept formation. Vinner & Dreyfus (1989) point out that usually the concept image is not build on definitions but essentially determined by typical examples:

> Hence, the set of mathematical objects considered by the student to be examples of the concept is not necessarily the same as the set of mathematical objects determined by the definition. (p. 356)

In traditional classroom settings, examples are primarily used to introduce a new concept. The learning and later reproduction of a concept is intertwined with diverse conscious as well as unconscious processes and the formed concept image will play a crucial role.

Teachers also expect a comparable one-way relationship from concept definition to concept image during problem solving. However, Vinner (1994) could show that the concept definition does not play any role when students are working on problems:

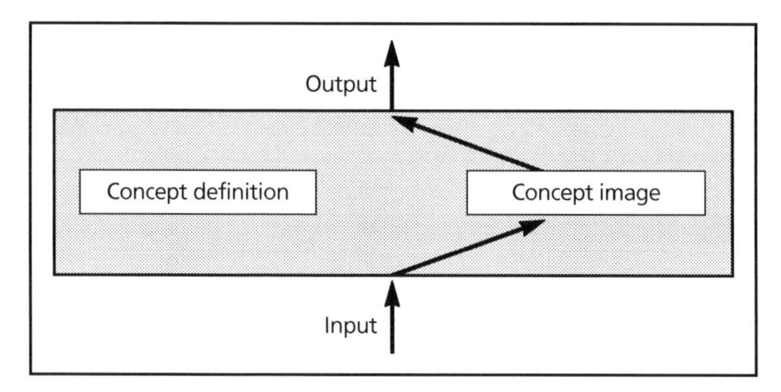

Figure 16.5 Significance of the concept image during problem solving (Vinner, 1994, p. 73).

When working on problems, students do not consider any concept definition. They base their decisions on the concept image. To that effect, Vinner (1994) could show how obstacles in calculus occurred since students remain on a restricted concept image of a tangent, already developed earlier. This concept is usually introduced as tangent on a circle. This concept image provokes difficulties in students' learning of calculus when confronted with a different, namely analytical, definition of a tangent. Additionally, elaborating on secant lines converging on a tangent line reinforces the hitherto developed concept image, which is mostly not enlarged by aspects of a more general tangent. Vinner (1994) found out that, among other things, students had difficulties to draw a tangent having more than one point in common with the curve.

The Integral Concept

According to the curricula for secondary schools in Germany, the central conceptualizations in calculus should be built on different intuitive ideas supporting each other. Nevertheless, these ideas should not be restricted to the area aspect but substantiated by also focusing on applications. The favored approach to integral calculus in German schools is linked to the Riemann integral without defining it explicitly. There are two main approaches to introduce the integral, the first one by calculating the area under a curve and the second one by approximation. Accordingly, in the traditional teaching in German classrooms a geometric and an analytic definition are introduced to students. The former refers to the oriented area of a region under a curve and the latter to the limit of a sum of areas of rectangles.

In the German didactical literature, these different approaches have been intensively discussed. We, however, restrict ourselves to refer to some basic works in this area. Already in 1976, Kirsch called for an "intellectual honest" introduction of this concept in order to promote appropriate learning of students. Tietze, Klika and Wolpers (1997) differentiate two approaches that focus on the area aspect and the antiderivative while Blum und Törner (1983) refer to the mean value as a third approach. Hußmann (2002) points to the usefulness of the approximation aspect by applying the idea of cumulation, which offers the advantage of easily interpreting negative values.

Nevertheless, teaching integral calculus in German schools is characterized by stable patterns. According to Blum (2000), it is mainly oriented at schemata and formulae. Consequently students are lacking a conceptual understanding of the integral. He underlines this observation by referring to the following TIMSS task, which could only be solved by 23% of German secondary school students (see Figure 16.6).

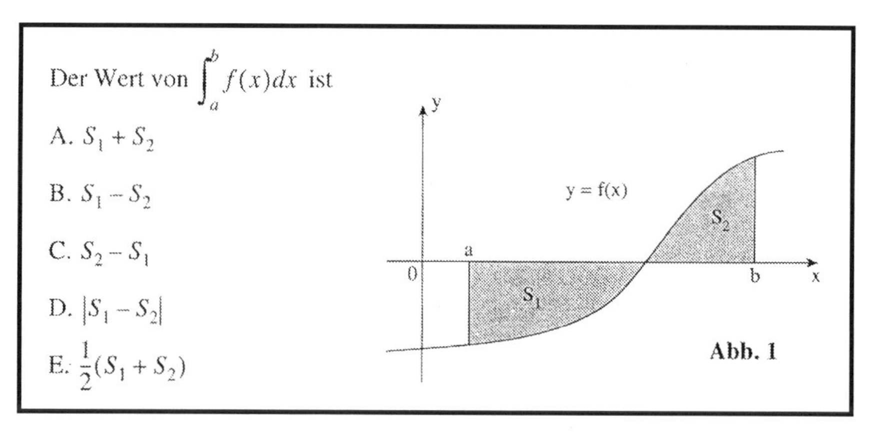

Figure 16.6 TIMSS task connecting area and integral (Blum, 2000).

The task shown in Figure 16.6 considers the oriented area aspect of the integral and the results showed that many students just equalled the concepts of integral and area. This tendency also emerged in our study which will be outlined in the results section.

METHODOLOGY

The design of the study is conceptually oriented at the study of Rasslan and Tall (2002). The participants in our study were students from grade 12 of one German secondary school. The class consisted of 24 students, 14 female and 10 male students. The students were enrolled in a so-called *Leistungskurs*, an advanced mathematics course with five hours per week compared to three hours in the general course.

In this section, we present the guiding research questions and the design of the study including the problems that we employed.

Research Questions

On the background of the aforementioned theoretical comments, we examined students' concept images and concept definitions of the integral. As mentioned before, our investigation was guided by a study from Rasslan and Tall (2002). Their underlying assumption is that letting students work on adequate problems provides insights into the concept images and concept definitions. The following research questions were central to our study:

- Which concept images and which concept definitions do students activate while working on problems around the integral?
- What alternative and individual concept images on the integral do students possess besides the intended ones?
- To what extent do the students reveal inconsistencies between the concept images and the concept definitions?

Design of the Study

In order to get various information on the concept definition and especially the concept image of the integral, we employed different methods for data collection—called "triangulation" by Cohen, Manion and Morrison (2007). Likewise, Schoenfeld (2002) suggests employing multiple data sources in order to enhance the trustworthiness of the data.

Although our study is conceptually oriented at the one of Rasslan and Tall (2002), it is modified with regard to the German classroom and refined to get deeper insight into students' concept formation concerning the integral. In comparison to five tasks on the concept image, there is only one question on the concept definition in the study of Rasslan and Tall (2002), *In your opinion what is*

$$\int_a^b f(x)dx$$

(the definite integral of the function f in the interval [a,b])? However, an analysis of German textbooks as well as lessons revealed that this question is too broad for the German situation (see also section 2.4). We therefore adopted a more sophisticated point of view and created three questions on the concept definition. Moreover, we explicitly asked the students not only to write down their solution but also to use sketches, to give explanations and to document their procedure. According to Rasslan and Tall (2002), we used five problems around the integral to get information about students' concept image. Furthermore, the concept image was investigated by inviting the students to draw a mind map. This is appropriated to represent in a graphic way the cognitive structure of a key topic including the related notions and the connections between them.

Mind Maps

Mind maps are considered as possibility to represent all aspects a person associates with a given concept. The name of the concept, "integral" in our case, serves as stimulus for recalling related contents and connections between them. These are organized in a diagram using lines to represent the relationships between different notions. A particularity of our task consists in the fact that the students were asked to enumerate the lines. This serves as information about the genesis of the mind map in the temporal order. Berger and Törner (2002) emphasize that this is a good means to get to know the central notions. In their opinion, the earlier a notion is recalled the more central it is to a person.

Beforehand, the students were shown via overhead projector an example of a mind map around the theme of function. Moreover, relevant aspects for the production of a mind map were recalled like those mentioned above.

Questionnaire

The questionnaire was employed in a pre-study, and a modified version was again implemented. Every time, the students had to work on the questionnaire in the classroom under supervision and were allowed to use a calculator.

Problems 1 to 3:

The concept definition was investigated by the following three problems:

Problem 1:

What do you understand by $\int_a^b f(x)dx$?

Problem 2:

Give a geometric definition of the integral and an illustration.

Problem 3:

Give an analytic definition of the integral and an illustration.

Problem 1 is meant to evoke students' associations with the symbolic notation of the integral. Problem 2 includes the area aspect while in problem 3, the approximation aspect is considered. With these three problems, we aimed at finding out which definition is familiar to the students and which aspects they remind in the respective context. As the students were asked to illustrate their solution by using sketches, the concrete conceptions on the integral can be checked. This includes, for example, the representation of the integral as positive or oriented area.

Problems 4 to 8:

The problems 4 to 8 are concerned with different aspects on the concept image of the integral.

Problem 4:

Find a formula for the area by using integration.

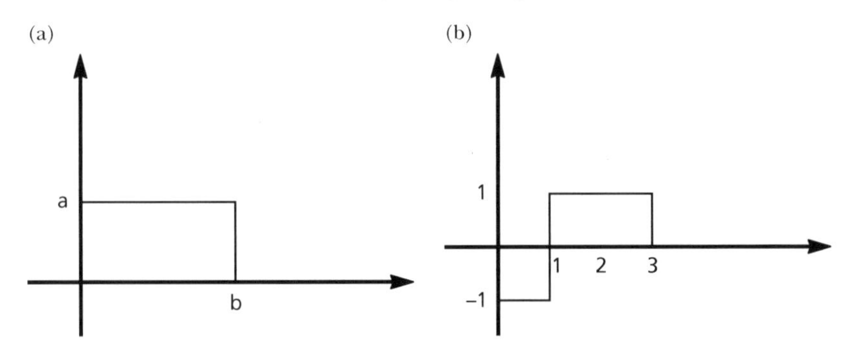

Problem 4 was designed to uncover if the students were able to use the integral for the calculation of simple areas. We hereby examined the naturalness of the students to manage the symbolic notation of the integral, especially their capacity to determine the boundaries of the integral and the integrand. In the first part of the task, the indication of the integral for a rectangle located above the x-axis is required. In the second part of the task, pieces of the area are situated in the 4th quadrant, that is, the integral here designates the oriented area.

Problem 5:

The picture shows two areas A and B. What do you think is correct for the relation between the areas?

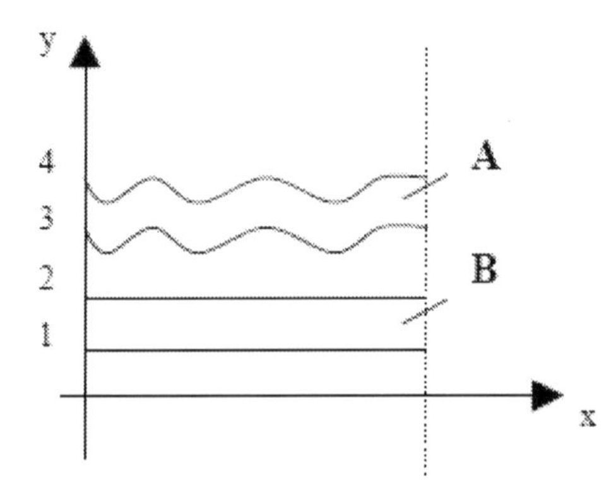

- The area of A is bigger than the one of B.
- The area of A is smaller than the one of B.
- Both areas are equal.
- Without any function given explicitly, it is not possible to answer this question.

Problem 5 is concerned with intuitive approaches to the solution in contrast to more logically oriented considerations. Vinner (1997) pointed out that "the concept image can be considered as part of intuition" and that "in some cases the intuitive mode of thinking just misleads us" (p. 67). In addition, a similar task to problem 5 can be found in a study from Fischbein (1999) who emphasizes its intuitive character:

> We posed this problem to high school students. The immediate reaction of these students was that the two areas are not equal and this means that there is nothing to prove. This was the *intuitive*, direct, apparently self-evident reaction to our problem. In reality, the two areas *are* equivalent, but this may be proven only indirectly by a logical analysis. (p. 20)

Problem 6:

a. Find the area bounded between the function f(x)=sinx and the x-axis over [π,2π].
b. Calculate the integral: $\int_{-\pi}^{2\pi} \sin x dx$.

In problem 6, conceptions on the area aspect of the integral are raised. Part a) asks for the calculation of the area that includes the graph of the sine function with the x-axis in the interval [π,2π]. In part b), the calculation of the integral of the sine function in the interval [–π,2π] is required. If the two tasks are considered in a holistic way, the illustration of the position of the area will facilitate the solution. A sketch, for example, would enable the students to answer problem 6b) immediately – assumed that they possess a profound understanding of the integral.

Problem 7:

How would you proceed to calculate the integral $\int_{-1}^{1} \sin(2x^3)dx$?

In problem 7, the competent handling of the integral by initially using visualization is tested. As the function is odd and point symmetric, its integral on the given interval [–1,1] equals zero (see Fig. 16.7). Compared to substitution or integration by parts, these simple and elementary considerations save time and laborious calculations.

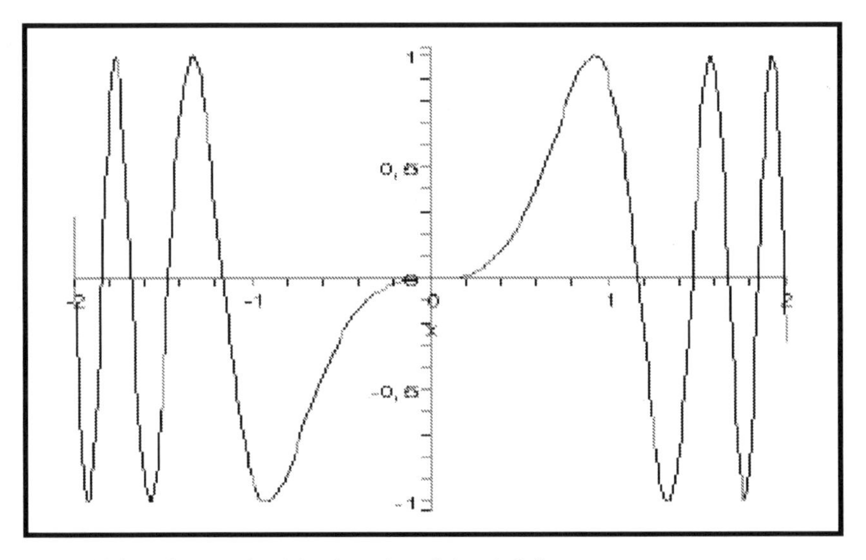

Figure 16.7 The graph of the function $f(x) = \sin 2x^3$.

Problem 8:

Give an example for a non-integrable function and explain your choice.

In problem 8, the students are asked to give an example for a function that is non-integrable and to explain their example. The underlying assumption is that the use of examples and counterexamples contributes considerably to the deep learning of concepts (Vollrath, 2001) and we wanted to know if counterexamples are also part of students' concept images.

RESULTS AND DISCUSSION

In this section, we present the results of our study. We first report on the problems that we designed to prompt the concept definition. However, if there become evident interesting aspects related to the concept image, we also mention these. We then present the results on the tasks explicitly created to evoke the concept image. Although we also used mind maps to investigate students' concept images we do not present these results in this paper and we therefore refer to Rösken (2004). To support our interpretations, we insert excerpts of students' answers.

Concept Definition

Problem 1:

All students worked on the first question. The responses reveal two main aspects. On the one hand, the students explain the symbol of the integral; on the other hand, they refer to the area aspect. Table 16.1 shows the distribution of the answers.

TABLE 16.1 Students' Answers to Problem 1

Students' answers	Number
The integral from a to b of the function f.	12
The area that the graph of f includes with the x-axis in the intervall $[a,b]$.	12
Others	3

The sum of the students in Table 16.1 does not equal 24 because three students mentioned both aspects. The twelve students that relate to the symbol of the integral the area aspect use quite vague formulations to express their ideas. The geometric conception associated with the integral is linked to simple area aspects. The following quotes from students' answers serve as examples to illustrate this:

- With this formula, you can calculate a certain area which is limited by the boundaries a and b.
- This is the area that the graph $f(x)$ includes between the interval $[a,b]$ and the x-axis.
- The graph $f(x)$ includes between $x = a$ and $x = b$ an area in relation to the x-axis which is determined through the integral.

The notion of oriented area is not mentioned by any student. Three students are not able to give an appropriate answer:

- It is the average value of $f(x)$ related to the domain from a to b.
- Hereunder, I understand the function $f(x)$ that describes the graph in the interval from a to b in relation to the variable x.
- This task deals with a function $f(x)$ that is located in the domain from a to b.

It is worth mentioning that in the students' answers the explication of the symbol of integral refers more to the concept definition while the interpretation as area is more related to the concept image. Finally, the results from Rasslan and Tall (2002) have already shown that this question is only in a restricted way appropriate to reveal representations on the concept definition.

Problem 2:

The analysis of the answers to this question is also guided by the formation of categories. The geometric definition of the integral relates to the geometric notion of area. As already mentioned, in German textbooks, the integral is interpreted as oriented area. The students' answers are put into categories in order to structure them according to their quality. First, we distinguished between three main categories. Within these, we further differentiate, as can be seen in Table 16.2:

TABLE 16.2 Students' Answers to Problem 2

Category	Content	Number
G1	**Appropriate definition**	**16**
G1a	Oriented area	1
G1b	Area for functions situated above the x-axis	15
G2	**Inappropriate definition**	**6**
G2a	Incorrect or incomplete definition	4
G2b	Only visual representation	2
G3	**No answer**	**2**

In order to clarify these categories, we further explain them and give examples from students' answers. In category G1a, we put the answers that refer to the formal concept definition. An important criterion of distinction to category G1b is the notion of oriented area. Figure 16.8 is an example to illustrate category G1a.

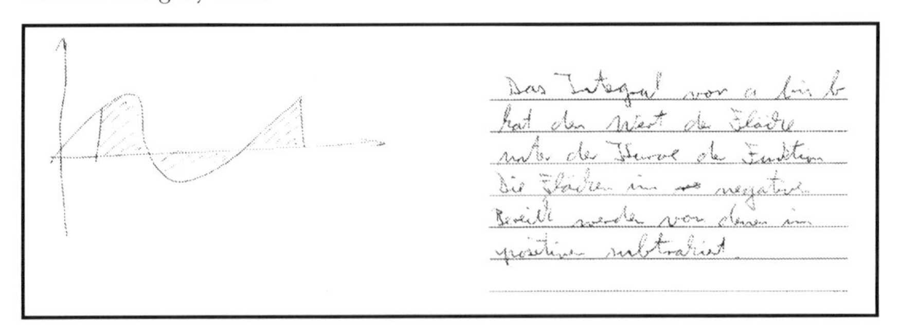

Figure 16.8 Example of a student's answer categorized as G1a.
[Translation: The integral from a to b has the value of the area under the curve of the function. The negative located areas will be subtracted from the positive ones.]

In category G1b, you can find the definitions that relate to an area of a function that is situated above the x-axis. Figure 16.9 serves as illustration of this category:

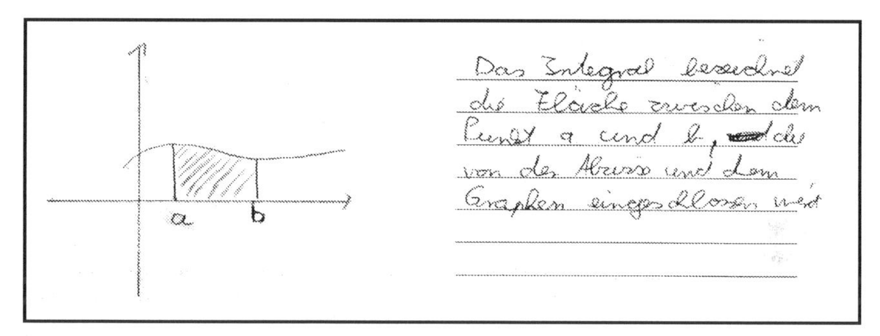

Figure 16.9 Example of a student's answer categorized as G1b.
[Translation: The integral indicates the area between the points a and b, which is included by the x-axis and the graph.]

In the two classrooms in which we employed this questionnaire, the introduction of the integral calculus was based on representations like the one illustrated in Figure 16.9 in order to motivate the area aspect. However, given the limited understanding that is expressed in answers like the one in Figure 16.9, the question arises whether these students have developed an adequate understanding of the integral. In the questionnaire, there are two problems that are concerned wit the interpretation of the integral as sum of positive and negative areas. The interesting question is, *How are students that possess a concept image in the sense of Figure 16.9 going to solve theses problems?*

Altogether, students' answers to problem 2 show that the area aspect is strongly linked to the image of a certain function, like the one sketched in Figure 16.9. This conception proves to be rather demonstrative and is remembered by the majority of the students.

Problem 3:

In the analytic definition, the area under a curve is approximated by polygons. This definition was initially introduced in the classroom for non-negative, continuous functions. Only later, this was concretized for the sum of positive and negative areas. For the interpretation of students' answers, we again distinguish between three main categories. The differentiation within these is more fine-grained than in question 2. This is due to the complexity of the analytic definition in comparison to the geometric one. Table 16.3 illustrates the categories and subcategories:

TABLE 16.3 Students' Answers to Problem 3

Category	Content	Number
A1	**Appropriate definition**	**6**
A2	**Inappropriate definition**	**18**
A2a	Good definitional attempts	9
A2b	Vague or absurd definitional attempts	8
A2c	No definition, only visual representation	1
A3	**No answer**	**6**

In nine students' answers, good definitional attempts become obvious. These students do know the constitutive aspects of the definition. However, it is difficult for them to express the connections appropriately. They describe the process of approximation but, at large, they are not able to define it adequately. A comparison between problem 2 to 3 shows that it is much easier for the students to define the integral using the geometric aspect. Yet, one has to take into account that this definition is restricted to the image of a specific function. The approximation by polygons turns out to be demonstrative as well and is used in 15 definitional attempts. However, the main ideas cannot be expressed as easily as in the case of the geometric definition of the integral.

Problem 4a:

Half of the students are able to give the correct result as shown in Table 16.4. More interesting and insightful is analyzing the incorrect answers in detail.

TABLE 16.4 Students' Answers to Problem 4a

Category	Content	Number
1	Correct result	12
2	Difficulties to name the integral	9
3	Calculation of the area without using the integral	3

Among the incorrect answers in category 2, the following terms can be found:

$$\int_a^b kdx, \int_a^b adx, \int_a^b a(x)dx, \int_a^b f(a)dx, \int_0^b ada.$$

One difficulty for the students is to name the limits of integration. It is evident that finding the integral for the given image conflicts with the standard notation. Furthermore, the students have major problems to recognize the given constant function as a possible integrand. Obviously, they are missing an x-term.

Figure 16.10 shows a solution that two students gave in the pre-study. Even if this was in the pre-study, we consider their procedure as worth mentioning here as well. These students solved the conflict mentioned above by drawing a supporting straight line as shown in Figure 16.10. Hence, they obtained the answer to this problem in a creative though complicated way.

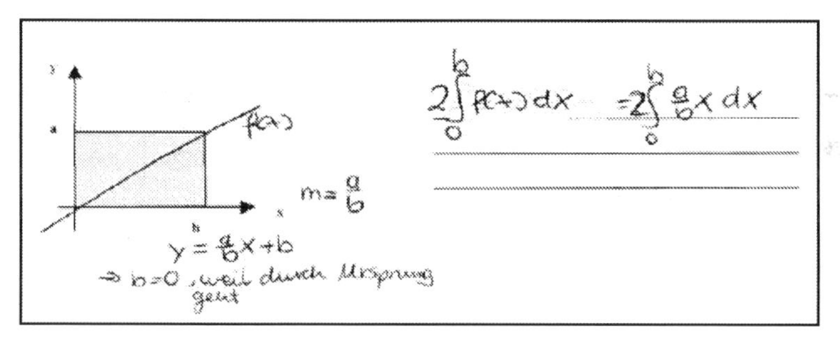

Figure 16.10 One student's solution to problem 4a.

This task documents that many students associate a certain image of function to the integral. In the case of the two students, they try to produce this concept image in order to be able to apply their knowledge.

Problem 4b:

As can be seen in Table 16.5, ten students gave the correct solution and 14 an incorrect one. Again, we have a closer look at the incorrect solutions:

TABLE 16.5 Students' Answers to Problem 4b

Category	Students' answers	Number	Aspect of the concept image
1	$\int_0^1 -1dx + \int_1^3 1dx$, the orientation of the area is not considered	4	Elementary conceptions of area
2	$\left\|\int_0^1 f(x)dx\right\| + \int_1^3 f(x)dx$, likewise $a(x)$, $f(1)$ or $f(-1)$ are named as integrands	7	Strong attachment to the traditional notion of integral
3	Elementary calculation of the area	2	—
4	$\int_0^3 (x-1)dx$, calculation of the area via a straight line	1	Calculation of the area is associated with a certain image of function

The answers show that the respective concept image of the students is linked to certain conceptions that constitute an obstacle for working on the task. In category 1, the students do not consider the position of the area while calculating it using the integral. Their concept image is bound to simple conceptions of area. In category 2, the students perfectly take into account the position of the areas, the mistake is due to another reason: they have difficulties to name the integrand. Their concept image for the symbol of the integral is associated to a certain type of function. In order to keep in line with this conception, these students indicate as integrand f(x), a(x) or f(1) and f(-1). One student gives the right solution but evaluates her result by the following statement:

> $\int_0^3 1dx$: Not possible, because this is a constant function and there is no x in it and that's why it is not possible to put in the limits.

It becomes evident here that the concept image does not exist in a consistent way. First, the student gives the correct answer but then notices that it is not in line with other parts of her conceptions. The conflict is solved by preferring the part of the concept image that is related to a specific function in the notion of the integral.

The students that we put in category 3 do not pay attention to the task and calculate the area without using the integral. They draw on elementary ways to calculate the area.

In category 4, the student tries to calculate the area by considering a "supporting straight line." He obtains the function $f(x) = x - 1$. Again, it becomes apparent that the concept image is associated to a certain image of function. The area under a constant function does not seem to correspond with this conception.

The two main categories, namely 1 and 2, do not appear jointly. This means that students who were put in category 1 indicate the right functions while students that were put in category 2 have difficulties to name the integrand but possess correct conceptions on the area. One student combines the elementary calculation of the area with the integral calculus by writing the following:

$$\int_0^1 a^2 dx + \int_1^3 abdx.$$

Summarized, while the difficulties to find the integrand remained, the problem to name the limits of integration minimized due to the concrete numbers provided in the illustration. However, a new obstacle emerged because of the orientation of the areas. Instead of the area, the students calculated the integral. Some students solved this conflict by shifting the square above the x-axis (see Figure 16.11).

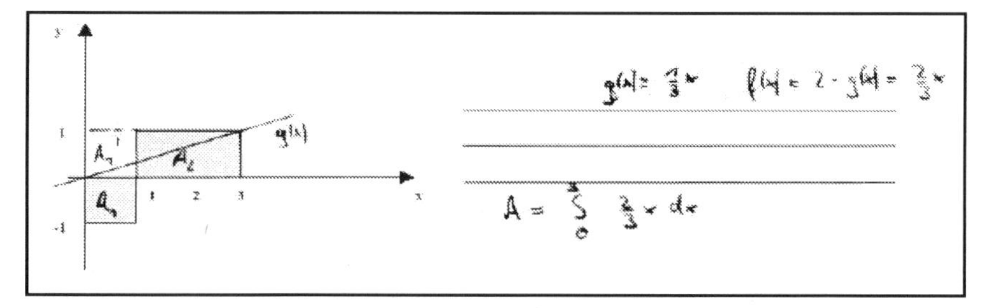

Figure 16.11 One example for students' solution to problem 4b.

Problem 5:

The areas A and B are equal. There are different possibilities to compare the two areas. 20 students give the right solution. Of these students, 17 explain this by stating that the respective functions have the same distance on the whole interval. Exemplarily, the following two answers document this:

- The areas are equal because the distances of the two limiting lines are equal.
- The x-section is the same, the y-section (that is between the limiting lines) as well, hence the areas are identical (for every x the difference of the limiting functions is identical).

One student refers to Cavalieri's principle. Two students draw an additional line and use it to show the equality of the areas. Figure 16.12 shows this solution.

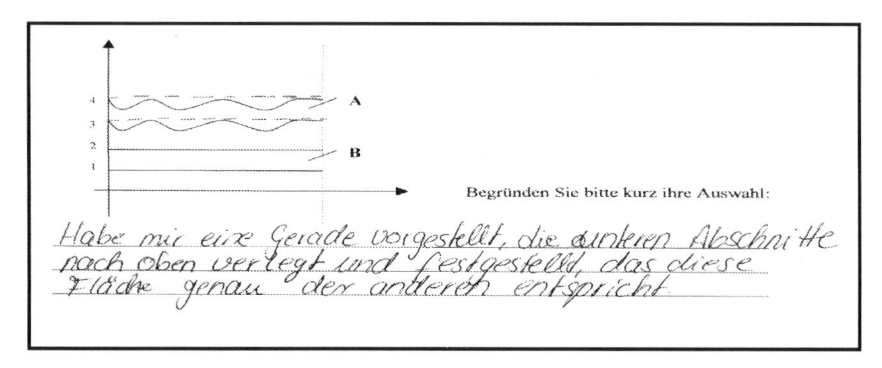

Figure 16.12 One student's solution to problem 5.
[Translation: I imagined a straight line, put the sections from below to above and noticed that this area just fits the other one.]

Two students argue that the area of A is bigger than the one of B. They give the following explanations:

- If I stretched out the area A, I would have two horizontal lines that would be longer than the lines of area B.
- As the two functions that include A, exactly as for B, always have the distance 1 but are waved and have the same limit, A > B.

Two students claim that an answer to this question is not possible because there is no function indicated. They state that—only by referring to the illustration—you cannot assume the parallelism of the functions that limit A. A student in the pre-study gives another interesting answer:

- Because of the curve respectively the waves, the graphs of A need as much as those of B. However, the graphs of A are much longer, hence, the area is also bigger than B. Example: The intestine in the body is much longer as one imagines but as it is curved and folded, it fits in our body.

It becomes obvious that the concept image does not only contain mathematical elements. The process of solving problems also involves everyday life experiences and considerations. All in all, the results show the importance of already existing knowledge and intuitive conceptions for the students' concept images. The concept images build on the experiences and knowledge that also might stem from other contexts. Spontaneous and intuitive considerations are assigned priority while analytical verifications are neglected.

Problem 6a:

The area that the students were to calculate is situated in the 4th quadrant. The area that the graph of $f(x)$ includes with the x-axis in the interval $[\pi,2\pi]$ equals 2. Ten of the 24 students are able to give the correct solution. The 14 incorrect solutions are described and analyzed in Table 16.6:

TABLE 16.6 Students' Answers to Problem 6a

Category	Students' answers	Number	Aspects of the concept image
1	–2 as result, i.e., negative area	7	integral = area
2	Incorrect antiderivative (cos x, –sin x)	5	—
3	Other arithmetic errors	2	—

Seven students calculate the area instead of the integral. Their answer results in giving a negative value as area of the function. On the one hand, these students do not use visualization to approach the problem. On the other hand, they do not scrutinize the negative value of their result. Some even add [FE] in order to indicate that they refer to the size of an area. In their concept image, the calculation of the integral is either equal to the one of the area or they do not consider it necessary to take into account the course of the graph. No student at all sketches the graph of the function.

Five of the seven students in categories 2 and 3 use the absolute value for the integral and document at least a correct procedure for the area aspect.

Problem 6b:

Considering the solution to problem 6a as well as the course of the graph, the value of the integral can immediately be given, namely –2. Here, nine students answer the problem correctly and 15 incorrectly. The analysis of the incorrect answers is similar to the one in problem 6a. This is why we abstain from presenting them here again. However, it is remarkable that some students continue to calculate the area instead of the integral as required in 6a. They give as answers a positive value, some of them even mention explicitly $A = 2$.

This task aimed at broaching the relationship between the integral and the related area. The results show that many students are not able to clearly distinguish between these two concepts. Their concept images equate the integral with the area. This relationship exists in both directions, as the two parts of the problem show.

Problem 7:

The integral equals zero. Eight students propose to work out the integral by finding the antiderivative, nine students by substitution and three students by integration by parts. Two students do not answer this question at all. Only two students take into account the course of the given function:

- Integration by substitution, using opposite boundaries of integration.
- The function is point symmetric to the point of origin; hence the integral (not the area) equals 0.

The first answer contains a vague allusion to the boundaries of integration. In the second answer, the student elaborates on this idea even further.

Already in the previous task, the sine function was subject of the considerations. Here, it is combined with another odd function. Therefore, the point symmetry of the graph is still obtained. The solutions to this task show an explicit bias towards an algorithmic approach even though the visual one would have been significantly easier. Aspects related to the calculation like, *How can I calculate the integral?*, come to the fore in students' considerations while relevant aspects of the integral take a back seat. They are captured in the procedure of calculation and fixed on algorithms instead of using the concept in a creative and flexible way. Interestingly, the mental images and conceptions that form the concept image as a whole are eclipsed. The introduction and the development of the integral are highly based on visual representations. However, these visual aspects of the integral do not seem to have the same relevance for the students while working on concrete problems. The starting point for students' reflections is not an appropriate visualization but rather oriented at the choice of an appropriate procedure.

Problem 8:

In this question, students were asked to indicate counterexamples that they knew or were able to create on the basis of their intuitive conceptions. Seven students mention as example of a non-integrable function a function with a pole:

- Spontaneously, only functions that have poles like the tangent function cross my mind. Those can be handled by considering limits.

In this context, six students refer to the function $f(x)=\frac{1}{x}$. The following statements can be found in the students' answers:

- Functions with poles: with poles, you cannot integrate $\int_{-1}^{1}\frac{1}{x}dx$.
- $\int_{-1}^{1}\frac{1}{x}dx$ because $\frac{1}{0}$ is not defined.
- $\int_{0}^{2}\frac{1}{x}dx$. This is non-integrable because 0 is not allowed for x, there is a pole.

These students justify their examples by pointing to the existence of a gap in the domain. Three students mention functions whose domain is the whole number line:

- x = IR (for example, $y = 1$, $y = 2$, $y = 3$, $y = 1.0006.783.986$). These functions are not integrable because in an arbitrary interval, they do not include any area.
- $\int f(x)dx$ because no intervals are given.
- $f(x) = 0$. As this function possess no area, it is not possible to integrate.

One female student connects the integrability with the differentiability:

- $f(x) = \text{sign}(x)$ non-differentiable $\rightarrow$ non-integrable.

Another interesting answer is given by a female student:

- $\int_{a}^{b} 5x \cdot 10x^2 \cdot 20x^3 \cdot 40x^4 dx.$

  We have only learned the product rule for two factors. Maybe there is a rule that allows for integrating this function as well but I do not know it.

The student connects the integrability with the existence of an antiderivative and mixes two different aspects. This idea is also hinted at in the following answer:

- I do not know it exactly, many [functions] come to my mind but I do not know if somebody could not integrate them in another way.

Only one student gives as an example the Dirichlet function.

The answers show that the concept image of the students contains various conceptions and explications to the notion of non-integrability that do not always describe the circumstances adequately.

CONCLUSION AND EDUCATIONAL IMPLICATIONS

The model of concept image and concept definition allows for analyzing the cognitive processes influencing the learning of mathematics by simultaneously considering mathematical characteristics. In general, understanding new information is essential in order to integrate new aspects into previously existing knowledge structures. Obviously, in learning mathematics students are to build rich and comprehensive concept images related to a specific notion. These concept images include own associations which make it possible for the students to grasp the ideas behind a mathematical notion. However, the formation of a concept image is an ambivalent process, in the sense that important aspects of the formal definitions are often not adequately represented. As a result, difficulties in students' learning might occur. The concept images could include critical aspects in comparison to the concept definitions.

In our empirical study we analyzed students' concept definitions and concept images of the definite integral as well as inconsistencies between them. In what follows, we sum up some of the main information we got out of the results to the problems. Our findings indicate that the students have developed rich concept images. The results also showed that most students knew the relevant aspects around the concept. As expected, the definitions were rather weakly represented, the findings were even worse for the analytic one. While working on the tasks to the integral many students ran into serious problems. The results raised the following conflicting issues of students' concept images:

- The concepts of area and integral are not distinguished.
- The symbol of integral is connected to a specific type of function.
- The calculation of the area is restricted to a specific graph of function.

It becomes apparent that the concept image has the character of a schema, including different ideas connected to the integral. If a task conflicts with the concept image, assimilation processes will occur. Parts of the task were ignored or interpreted in a different way. Further, we could see that the ideas of students cannot only be explained by their current concept image but ideas on other mathematical objects, too. We therefore conclude that a concept image cannot be seen as a single unit but is integrated in a networking of different concept images with manifold relations between them (see Fig. 16.13).

This networking aspect became obvious because of the difficulties students encountered when working on problem 4. Particularly, their evoked concept image of a function affected the problem solving process. That is, the flexibility in dealing with the current concept is restricted by difficulties related to other topics. The integration of knowledge is a sensitive process, problems in application situations easily occur.

Intuitive ideas are an important and decisive part of a concept image, as could be seen in the results to problem 5. Not only mathematical knowledge but also other experiences have a considerable effect on problem solving. Intuitive associations occur spontaneously, an analytic verification is often

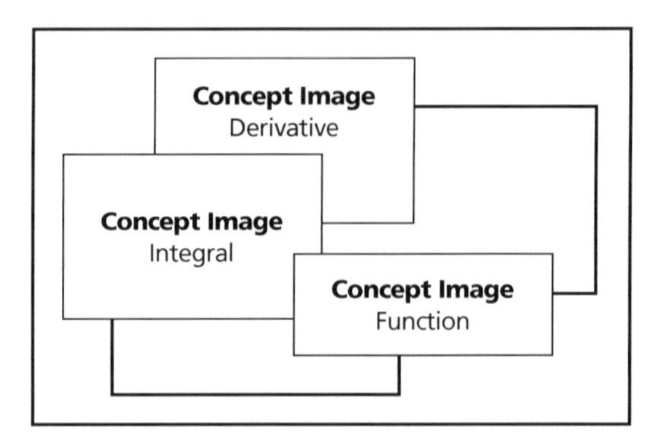

Figure 16.13 Networking of concept images related to different mathematical notions.

neglected. In most cases, these intuitive ideas dominate the concept image and often contain critical aspects compared to the concept definition.

In problem 6, the connection between the idea of an integral and the idea of an area was raised. As already could be concluded from students' answers to the geometric definition, the relation between the two notions is restricted to just one aspect. When working on the problems, these misconceptions became even more apparent. Moreover, some students are guided by the principle of always employing absolute value when calculating an area by using integration. For many students this rule is part of their concept image, and they are therefore not able to clearly distinguish between the concepts of area and integral. Furthermore, students did not consider it necessary to think about the course of the graph.

Another interesting aspect emerged from the results to problem 7. The evoked concept image of students was primarily restricted to techniques and procedures to the disadvantage of visualization, for example. That is, more creative approaches are not applied although they would be helpful. This focus on algorithms and calculations can also be seen in the other problems. We concentrated on this aspect in another paper in more detail (Rösken & Rolka, 2006). When we asked students for a non-integrable function in problem 8, a lot of connections to other mathematical topics occurred. What became apparent was that students had problems to link ideas from different contexts and to bring together different concept images. Without doubt, students possess a lot of conceptual knowledge as could be seen in the results.

Concerning the methodological approach, we conclude that we got wide-ranging information on students' concept images and concept definitions due to applying different methods. This study is part of ongoing research and in a follow-up study we are going to improve the following issues. First, we will additionally conduct interviews with the students in order to get deeper insight into their approaches, difficulties and subjective concept formations. Second, we will discuss some of the eye-catching students' results with the teachers to sensitize them for their students' intuitive approaches.

Finally, we can say that Tall and Vinner (1981) provide a powerful model to deeper analyze the construction of mathematical knowledge. Related to classroom praxis, the following questions can be raised, *How can students be inspired to build rich and useful concept images? What should teachers know about students' corresponding cognitive processes?* The critical aspects of the concept image that have been described in this paper could also offer a possibility to make the underlying conflict transparent for both students and teachers. Essential is to help students integrating intuition but unfortunately, many teachers are—for different and understandable reasons—primarily concerned with judging and evaluating "incorrect" solutions. These are rashly labelled as misconceptions instead of using their inherent potential for teaching.

REFERENCES

Bauer, L. (1995). Objektive mathematische Stoffstruktur und Subjektivität des Mathematiklernens. In H.-G. Steiner & H.J. Vollrath (Hrsg.), *Untersuchungen zum Mathematikunterricht, Band 20: Neue problem- und praxisbezogene Forschungsansätze* (pp. 9–16). Köln: Aulis.

Berger, P. & Törner, G. (2002). Some Characteristics of Mental Representations of Exponential Functions—A Case Study with Prospective Teachers. In *Schriftenreihe des Instituts für Mathematik der Gerhardt Mercator Universität Duisburg,* Nr. 539. Duisburg: Universität Duisburg.

Blum, W. (2000). Perspektiven für den Analysisunterricht. *Der Mathematikunterricht, 46*(4/5), 5–17.

Blum, W. & Törner, G. (1983). *Didaktik der Analysis.* Göttingen: Vandenhoeck & Ruprecht.

Cohen, L., Manion, & Morrison, K. (2007). *Research Methods in Education* (6th ed.). London: Routledge.

Cornu, B. (1981). Apprentissage de la Notion de Limite: Modèles Spontanés et Modèles Propres. In C. Comiti & G. Vergnaud (Eds.), *Proceedings of the fifth conference of the International Group* (S. 322–326). Grenoble: University.

Even, R. & Tirosh, D. (2002). Teacher Knowledge and Understanding of Students' Mathematical Learning. In L. English (Ed.), *Handbook of International Research in Mathematics Education* (pp. 219–240). Mahwah, NJ: Erlbaum.

Fischbein, E. (1987). Intuition in Science and Mathematics. An Educational Approach. Dordrecht: Kluwer.

Freudenthal, H. (1973). Mathematik als pädagogische Aufgabe (Band 2). Stuttgart: Klett.

Harel, G. & Sowder, L. (2005). Advanced Mathematical-Thinking at Any Age: Its Nature and Its Development. *Mathematical Thinking and Learning,* 7 (1), 27–50.

Hartland, J. (1995). Sprache und Denken. In: P. Banyard (Hrsg.), *Einführung in die Kognitionspsychologie* (S. 195–244). München: Ernst Reinhardt.

Hußmann, S. (2002). Konstruktivistisches Lernen an Intentionalen Problemen. Mathematik unterrichten in einer offenen Lernumgebung. Hildesheim, Berlin: Franzbecker.

Kirsch, A. (1976). Eine "intellektuell ehrliche" Einführung des Integralbegriffs in Grundkursen. *Didaktik der Mathematik 4,* S. 97–105

Köller, O., Baumer, J. & Neubrand, J. (2000). Epistemologische Überzeugungen und Fachverständnis im Mathematik- und Physikunterricht. In J. Baumert, W. Bos & R. Lehmann (Hrsg.), *TIMSS/III Dritte Internationale Mathematik- und Naturwissenschaftsstudie: Mathematische und naturwissenschaftliche Bildung am Ende der Schullaufbahn, Band 2: Mathematische und physikalische Kompetenzen am Ende der gymnasialen Oberstufe* (pp. 229–269).

Ouvrier-Buffet, C. (2006). Exploring Mathematical Definition Construction Processes. *Educational Studies in Mathematics, 63* (3), 259–282.

Poincaré, H. (1908/52) . Science and Method. New York: Dover. (Original published in 1908: Science et méthode).

Presmeg, N. (2006). Research on visualization in learning and teaching mathematics. In A. Gutiérrez, p. Boero (Eds.), *Handbook of Research on the Psychology of Mathematics Education: Past, Present and Future* (pp. 205–235). Rotterdam, Taipei: Sense Publishers.

Rasslan, S. & Tall, D. (2002). Definitions and images for the definite integral concept. In Anne D. Cockburn & Elena Nardi (Eds), *Proceedings of the 26th Conference of the International Group for the Psychology of Mathematics Education,* (Norwich, UK), 4, 89–96.

Rösken. B. (2004). *Empirische Erhebung zur Analyse der mentalen Repräsentation von Concept Imgage und Concept Definition.* (Staatsarbeit, not published).

Rösken, B. & Rolka, K. (2006). A picture is worth a 1000 words—The role of visualization in mathematics leariring. In J. Novotná, H. Moraová, M. Krátká & N. Stehlíková (Eds.), *Proceedings of the 30th Conference of the International Group for the Psychology of Mathematics Education* (Vol. 3, pp. 233–240). Prague, Czech Republic: PME.

Rosch, E. (1975) . Cognitive reference points. *Cognitive Psychology* 7, 532–547.

Schoenfeld, A. (1998). Toward a theory of teaching-in-context. *Issues in Education, 4*(1), 1–94.

Schoenfeld, A. H. (2002). Research methods in (mathematics) education. In L. English (Ed.), *Handbook of International Research in Mathematics Education* (pp. 435–488). Mahwah, NJ: Erlbaum.

Tall, D. (1994). The psychology of advanced mathematical thinking: Biological brain and mathematical mind. In J. P. Ponte & J. F. Matos (Eds.), *Proceedings of the 18th Conference of the International Group for the Psychology of Mathematics Education* (Vol. 1, pp. 33–39). Lisbon, Portugal: PME.

Tall, D., & Vinner, S. (1981). Concept image and concept definition in mathematics with particular reference to limits and continuity. *Educational Studies in Mathematics, 12*(2), 151–169.

Tietze, U.-P., Klika, M. & Wolpers, H. (1997). *Mathematikunterricht in der Sekundarstufe II. Band 1: Fachdidaktische Grundfragen—Didaktik der Analysis.* Braunschweig: Vieweg.

Tversky, A. & Kahneman, D. (1974). Judgement under uncertainty: Heuristics and biases. *Science, 185,* 1124–1130.

Vinner, S. (1994). Research in teaching and learning mathematics at an advanced level. In D. Tall (Ed.), *Advanced Mathematical Thinking (2nd ed.).* Dordrecht: Kluwer.

Vinner, S. (1997). From intuition to inhibition—Mathematics, education and other endangered species. In E. Pehkonen (Ed.), *Proceedings of the 21st Conference of the International Group for the Psychology of Mathematics Education* (Vol. 1, pp. 63–78). Helsinki, Finland: PME.

Vinner, S. & Dreyfus, T. (1989). Images and Definitions for the Concept of Function. *Journal for Research in Mathematics Education 20* (4), 356–366.

Vinner, S. & Hershkowitz, R. (1980). Concept images and common cognitive paths in the development of some simple geometrical concepts. In R. Karplus (Ed.), *Proceedings of the International Conference for the Psychology of Mathematics Education* (pp. 177–184). Berkeley, California: University.

Vollrath, H.-J. (2001). *Grundlagen des Mathematikunterrichts in der Sekundarstufe.* Heidelberg, Berlin: Spektrum.

9 781593 118686